Environmental Sustainability of Current Waste Management Practices

Environmental Sustainability of Current Waste Management Practices

Editors

Rita Khanna
Yury Konyukhov
Igor Burmistrov

MDPI • Basel • Beijing • Wuhan • Barcelona • Belgrade • Manchester • Tokyo • Cluj • Tianjin

Editors

Rita Khanna
University of New South Wales (UNSW) Australia
Australia

Yury Konyukhov
National University of Science and Technology "MISiS"
Russia

Igor Burmistrov
Plekhanov Russian University of Economics
Russia

Editorial Office
MDPI
St. Alban-Anlage 66
4052 Basel, Switzerland

This is a reprint of articles from the Special Issue published online in the open access journal *Sustainability* (ISSN 2071-1050) (available at: https://www.mdpi.com/journal/sustainability/special_issues/waste_practices).

For citation purposes, cite each article independently as indicated on the article page online and as indicated below:

LastName, A.A.; LastName, B.B.; LastName, C.C. Article Title. *Journal Name* **Year**, *Volume Number*, Page Range.

ISBN 978-3-0365-3449-7 (Hbk)
ISBN 978-3-0365-3450-3 (PDF)

Contents

sustainability

MDPI

Editorial

Environmental Sustainability of Current Waste Management Practices

Rita Khanna [1,*], Yuri Konyukhov [2] and Igor Burmistrov [3]

1 School of Materials Science and Engineering (Ret.), The University of New South Wales, Sydney, NSW 2052, Australia
2 Department of Functional Nanosystems and High-Temperature Materials, National University of Science and Technology "MISiS", 119049 Moscow, Russia; ykonukhov@misis.ru
3 Engineering Centre, Plekhanov Russian University of Economics, 117997 Moscow, Russia; burmistov.in@rea.ru
* Correspondence: rita.khanna66@gmail.com

The Special Issue on '*Environmental Sustainability of Current Waste Management Practices*' was a part of the section '*Environmental Sustainability and Applications*' of the journal '***Sustainability***'. It was focused on recent developments in the field of 'Waste Management', including solid and liquid wastes and their collection, segregation, disposal and processing in a cost-effective and environmentally sustainable manner. Reports on different types of wastes included municipal, industrial, medical, construction, demolition, agricultural, electronic, hazardous, sewage sludge, etc. The focus was on mitigating various environmental issues and on the steps taken to enhance the sustainability of current waste management practices around the globe.

This 'Special Issue book' has six full articles with contributing authors from Australia, Chile, India, Israel, Kazakhstan, Nigeria, the Russian federation and Saudi Arabia. Global authorship in this book reflects multifaceted interest and activity in this field worldwide, with breakthroughs occurring in several research areas. Brief summaries and key features of various articles are given next.

The first article by Artyukhov et al. on 'Harvesting waste thermal energy using a surface-modified carbon fiber based thermo-electrochemical cell' focussed on the conversion of waste heat into electrical energy using thermo-electrochemical (TEC) cells. The utilization of energy from primary energy sources to their final use is accompanied by several losses in the form of waste heats; nearly 72% of the primary energy consumed can be lost as waste heat. This article presents new results on enhancing the efficiency of TEC cells based on carbon fiber electrodes and potassium ferri-/ferrocyanide redox electrolyte. Electrode surfaces were modified using magnetron deposition of silver and titanium and/or infiltration implantation of nanoscale titanium oxide. Surface modification of electrodes were found to change the internal resistance of TEC cells by three orders of magnitudes. Maximum power achieved with modified electrodes was determined to be 25 mW/m^2 and an efficiency of 1.37%.

The second article by Alharbi et al. on 'Toward sustainable environmental management of healthcare waste: A holistic perspective' presents a case study on healthcare waste-management practices in Saudi Arabia. A multi-faceted approach involving policy analysis, observations, semi-structured interviews and focus groups were used to elucidate the basics of healthcare waste management. It was estimated that Saudi government hospitals across the country discarded several waste items such as paper (27,000 tons), plastics (15,000 tons), food (10,000 tons), glass (8000 tons), and metal (7000 tons) in landfills every year with negligible levels of recycling. The lack of legal frameworks, waste-management training, coordination among stakeholders, and the lack of social responsibility were identified as some of the key challenges facing the system.

Citation: Khanna, R.; Konyukhov, Y.; Burmistrov, I. Environmental Sustainability of Current Waste Management Practices. *Sustainability* **2022**, *14*, 2321. https://doi.org/10.3390/su14042321

Received: 14 February 2022
Accepted: 17 February 2022
Published: 18 February 2022

The third article by Cayumil et al., 'An overview on solid waste generation and management: Current status in Chile', presented an overview on municipal solid wastes (MSW) and industrial waste from the iron/steelmaking and aluminium industries. Key waste issues such as sources, compositions, volumes, factors affecting waste generation and waste processing were first discussed, followed by further discussion on recycling, resource recovery, disposal and environmental impacts. Waste generation and management in Chile was presented in greater detail as a special case study. For Chile, being the world's largest producer of copper, with significant efforts for mining waste management, its infrastructure and procedures were presented to reduce the environmental impact of the mining sector and associated waste generation. Government initiatives, legislation for integrated solid waste management and measures were presented including regulations on waste management frameworks concerning transboundary movements of hazardous wastes, persistent organic pollutants, the closure of mining activities and installations, restrictions on plastics disposal, etc.

The fourth article by Ibikunle et al. on the 'Development of a software system for selecting steam power plant to convert municipal solid waste to energy' has reported on a thermodynamic-based software for the combustion of MSW into energy in a steam power plant with specific focus on the amounts of waste converted, heating values and capacities of power plants. Using 584 tons of MSW and a heating value of 20 MJ/kg as the input, an algorithm (Jawa script) computed saturated and superheated steam tables along with the requisite thermodynamic behaviour of the power plant operation. The software predicted a 3245 MWh energy potential for the quantity of waste investigated, with an electrical power potential of 41 MW. Plant capacities included 100 MW of boiler power, 41 MW of turbine power, and 60 MW of condenser power. This technique is expected to be a valuable tool in the waste-to-energy sector for processing MSW and power generation.

The fifth article by Dosmukhamedov et al. on 'Processing dross from hot-dip galvanizing by chlorination roasting' reports on recovering pure zinc ingots and zinc oxide from waste dross from hot-dip galvanizing towards utilization as mineral additives in animal and poultry feeds. The influence of chlorinating reagents $CaCl_2$ and NH_4Cl on the roasting temperature and degree of sublimation of Pb, Fe, Ni, Cu and Cd was investigated. The best results were obtained by using blends of $CaCl_2$ (6%) and NH_4Cl (15%) in proportion to the weight of the feed material. Optimal roasting parameters were identified as: 1000 °C, 60 min, and 0.1 L/min air flow. Impurity levels in pure zinc were determined to be: 0.05 Pb, 0.15 Fe, 0.06 Ni, 0.003 Cu and 0.001 Cd. The degree of sublimation achieved for copper, nickel and iron chlorides was ~75%, with lead and cadmium removal reaching 90–98% of their initial amount in the dross.

The sixth article by Khanna et al. on 'Red mud as a secondary resource of low-grade iron: A global perspective' assessed the suitability of red mud (RM) as a low-grade iron resource. Managing RM, a solid waste byproduct of the alumina recovery process, is a serious ecological and environmental issue. With ~150 million tonnes/year of RM being generated globally, nearly 4.6 billion tonnes of RM are presently stored in vast waste reserves. RM can be a valuable resource of metals, minor elements and rare earth elements. The utilization of RM as a material resource in several commercial and industrial operations was briefly reviewed along with key features of iron recovery techniques. RMs from different parts of the globe including India, China, Greece, Italy, France, Russia were examined for their iron recovery potential. The composition range of RMs examined were: Fe_2O_3: 28.3–63.2 wt.%; Al_2O_3: 6.9–26.53 wt.%; SiO_2: 2.3–22.0 wt.%; Na_2O: 0.27–13.44 wt.%; CaO: 0.26–23.8 wt.%; Al_2O_3/SiO_2: 0.3–4.6. Even with a high alumina content and high Al_2O_3/SiO_2 ratios, it was possible to recover metallic iron in all cases, showing the significant potential of RM as a secondary resource of low-grade iron.

This book covers extensive areas of interest on waste processing, recycling, material recovery, and the environmental impact of sustainable waste management, along with recent developments in the field. This book has a global perspective and wide coverage of topics for academics, professionals, regional and international organizations.

Author Contributions: R.K., Y.K. and I.B. contributed equally to writing this editorial. All authors have read and agreed to the published version of the manuscript.

Funding: This research received no external funding.

Conflicts of Interest: Authors declare no conflict of interest.

sustainability

MDPI

Article

Harvesting Waste Thermal Energy Using a Surface-Modified Carbon Fiber-Based Thermo-Electrochemical Cell

Denis Artyukhov [1], Nikolay Kiselev [1,2,3], Nikolay Gorshkov [1], Natalya Kovyneva [1], Olga Ganzha [1], Maria Vikulova [1], Alexander Gorokhovsky [1], Peter Offor [4,5], Elena Boychenko [2,3] and Igor Burmistrov [1,2,3,*]

1 Department of Chemistry and Chemical Technology of Materials, Yuri Gagarin State Technical University of Saratov, 410054 Saratov, Russia; mr.tokve@gmail.com (D.A.); nikokisely12345@gmail.com (N.K.); gorshkov.sstu@gmail.com (N.G.); k.natasha_86@bk.ru (N.K.); gangaolya@gmail.com (O.G.); vikulovama@yandex.ru (M.V.); algo54@mail.ru (A.G.)
2 Department of Functional Nanosystems and High-Temperature Materials, National University of Science and Technology «MISiS», 119049 Moscow, Russia; elena.boychenko.sar@gmail.com
3 Engineering Center, Plekhanov Russian University of Economics, 117997 Moscow, Russia
4 Metallurgical and Materials Engineering Department, University of Nigeria, Nsukka 410001, Nigeria; ykonukhov@misis.ru
5 Africa Centre of Excellence for Sustainable Power and Energy Development (ACE-SPED), University of Nigeria, Nsukka 410001, Nigeria
* Correspondence: burmistrov.in@misis.ru

Abstract: An important direction in the development of energy saving policy is harvesting and conversion into electricity of low-grade waste heat. The present paper is devoted to the improvement of the efficiency of thermo-electrochemical cells based on carbon fiber electrodes and potassium ferri-/ferrocyanide redox electrolyte. The influence of the carbon fiber electrode surface modification (magnetron deposition of silver and titanium or infiltration implantation of nanoscale titanium oxide) on the output power and parameters of the impedance equivalent scheme of a thermo-electrochemical cell has been studied. Two kinds of cell designs (a conventional electrochemical cell with a salt bridge and a coin cell-type body) were investigated. It was found that the nature of the surface modification of electrodes can change the internal resistance of the cell by three orders of magnitude. The dependence of the equivalent scheme parameters and output power density of the thermoelectric cell on the type of electrode materials was presented. It was observed that the maximum power for carbon fiber modified with titanium metal and titanium oxide was 25.2 mW/m^2 and the efficiency was 1.37%.

Keywords: thermo-electrochemical cell; waste heat harvesting; carbon fiber; surface modification; efficiency

Citation: Artyukhov, D.; Kiselev, N.; Gorshkov, N.; Kovyneva, N.; Ganzha, O.; Vikulova, M.; Gorokhovsky, A.; Offor, P.; Boychenko, E.; Burmistrov, I. Harvesting Waste Thermal Energy Using a Surface-Modified Carbon Fiber-Based Thermo-Electrochemical Cell. *Sustainability* **2021**, *13*, 1377. https://doi.org/10.3390/su13031377

Academic Editor: Jose Navarro Pedreño
Received: 8 December 2020
Accepted: 25 January 2021
Published: 28 January 2021

1. Introduction

The attention of the scientific community has recently been focused on generating the cheapest and cleanest energy. There are many areas where the amount of energy dissipated in the atmosphere is enormously high. This is especially significant in places where heat radiators and pipes with hot liquid are used. Moreover, a lot of potential heat is generated by mechanical and electrical equipment.

The harvest of low-temperature waste heat is a widely studied topic nowadays. Thermo-electrochemical cells (thermocells or thermo-galvanic cells, (TECs)) are one of the cheapest means to convert low-grade waste heat into electricity [1–4]. Typically, a TEC is based on a redox electrolyte and two electrodes placed at different temperatures [5–12]. The emergence of the temperature difference between the electrodes, in other words, the entropy difference between the two sides of the redox process, generates a potential difference and maintains a continuous flow of electric current in the TEC.

The conversion process is based on the temperature coefficient of the electrode potential or hypothetical Seebeck coefficient of the cells. The output voltage is proportional to the temperature difference between hot and cold electrodes because of the difference in entropy of the redox process for hot and cold electrodes [2,3]. The hypothetical Seebeck coefficient of thermo-electrochemical cells can be calculated by the following formula [4]:

$$Se = \frac{\delta V}{\delta T} = -\frac{\Delta S^0_{rx}}{nF} \tag{1}$$

where δV is the full-cell voltage, δT is the interelectrode temperature difference, ΔS^0_{rx} is the entropy change for the cell reaction, n is the number of electrons transferred in the reaction, and F is Faraday's constant.

Currently, researchers are focused on the $[Fe(CN)_6]^{3-}/[Fe(CN)_6]^{4-}$ redox couple and carbon nanomaterials with high specific surface areas (carbon nanotubes and reduced graphene oxide) as they exhibit fast electron transfer kinetics and a hypothetical Seebeck coefficient of about 1.4 mV/K. The highest reported Seebeck coefficient of 9.9 mV/K has been demonstrated for a thermocell containing acetone and iso-propanol as the redox couple, when the vaporization entropy of acetone increases the total entropy change in the conversion of iso-propanol to acetone [13]. The highest hypothetical Seebeck coefficient for aqueous electrolyte-based thermocells (4.5 mV/K) and, accordingly, high open-circuit voltage values of up to 0.2 V have been presented in studies [14,15].

The accumulation and storage of charge in electrochemical systems (capacitors, thermo-electrochemical cells) occur in the electric double-layer at the electrode/electrolyte junction. Therefore, electrode materials must meet the following requirements: high electrical conductivity, high specific surface area, availability of the porous structure to ions and electrolyte molecules, and low density. Carbon materials satisfy these requirements [16–20]. The relevance of carbon materials for electrochemical applications is due to the unique combination of chemical and physical properties: carbon electrodes are well polarized, stable over a wide range of temperatures, and chemically inert. In addition, they are characterized by high conductivity, high specific surface area, and relatively low cost and have a porous structure which can be controlled.

The best world records of TEC efficiency and power output were achieved with the catalytic modification of carbon nanotube aerogel by Pt nanoparticles [1]. In this work, a maximum power output of 6.6 W m^2 and a Carnot relative efficiency of 3.95% described in the literature have been shown. Several studies have described that the addition of metal nanoparticles provides a significant catalytic effect and increases the current density in a continuous redox reaction in a system of $[Fe(CN)_6]^{3-}/[Fe(CN)_6]^{4-}$ [4,21,22].

Nevertheless, carbon nanotubes are still limited in industrial production, are extremely expensive, and require the use of polymer binders for the creation of electrode materials. Simultaneously, carbon fiber (CF) materials have a significant advantage because they do not require the addition of binding components (in comparison with polymer additives for CNTs and carbon black-based electrodes [23,24]) in the manufacture of electrodes. CF materials are obtained mainly on the basis of polyacrylonitrile (PAN) pitch and rayon-based precursor. Isotropic pitch and rayon-based carbon fibers are excellent materials for the production of activated carbon fibers with very high specific areas (>1500 m^2/g) and are likely to dominate as materials for liquid as well as gaseous adsorption and environmental protection.

Therefore, in the present work, materials based on commercially available carbon fiber modified by easily reproducible magnetron sputtering (with Ti and Ag coating) and infiltration methods (with TiO_2 nanopowder) were investigated as cheap electrode materials for thermo-electrochemical cells based on the $[Fe(CN)_6]^{3-}/[Fe(CN)_6]^{4-}$ redox system, which can be easily scaled.

2. Materials and Methods

2.1. Materials

For electrolyte preparation, $K_3[Fe(CN)_6]$ (99.8% purity, CAS 13746-66-2) and $K_4[Fe(CN)_6]$ (99.8% purity, CAS 14459-95-1) were purchased from Sigma-Aldrich GmbH (Schnelldorf, Germany) and used without further purification. Ferri-/ferrocyanide redox electrolyte in the ratio of 1:1 with a total concentration of 0.3 mol/L in distilled water was used as the base.

Carbon fiber Busofit TM-04 (Busofit) commercial products were purchased from Svetlogorsk Khimvolokno OJSC (Svetlogorsk, Republic of Belarus), with a diameter of the initial fiber of about 6–8 μm and an electrical resistivity of approximately 20–200 Ω m.

Four modifications of Busofit carbon fabric were used as electrodes. Modification was carried out by vacuum sputtering of titanium using magnetron sources and condensing its vapors onto a carbon cloth tape:

1. Basic Busofit;
2. Busofit with Ag magnetron sputtering;
3. Busofit with Ti magnetron sputtering;
4. Busofit with infiltrated dispersion of TiO_2 nanoparticles.

The dispersal of titanium oxide (from a 5% dispersion of TiO_2) into Busofit fabric was carried out in vacuum, in a Buchner filter on a Bunsen funnel. Then, the coating was dried for 1 h at temperatures of 80 and 150 °C.

For cells in the coin cell CR2025, an industrial separator for Li-ion current sources was used.

2.2. Measurement and Characterization Technique

The potentiodynamic method was used to obtain the maximum power values. To determine the voltage of the open circuit of the thermal cell on the cold and hot sides, the following temperature differences were consistently created: 10, 20, 30, 40, and 50 K. After some time, a stable potential was established in the system, and the value of this potential was taken as the open circuit voltage (U_{OC}), which is unique to each material and each temperature. After that, the cell was discharged by creating a reverse potential, increasing from 0 mV to the open circuit voltage (U_{OC}, which was established in the previous step). U_{OC} was increased at the rate of 1 mV/s. During all experiments, the current flowing through the cell was recorded.

One of the stages of processing the measurement results was the construction and analysis of an equivalent scheme (ES), i.e., electrical circuits having the same frequency dependence on impedance as the studying cell. It can be used to determine the structural features of the investigated systems (electrode–electrolyte heterojunctions, particle coupling in clusters, surface and bulk conductivities), microscopic parameters (cluster sizes and their local resistance), and other characteristics of the objects under study.

A typical ES consists of several different elements: resistors, capacitors, inductors, Warburg impedance, and constant phase shift element [8]. Analysis of the ES and calculation of the total resistance are associated with the addition of harmonic oscillations of currents and voltages. In case of studying the dependence of the output parameters on the presence and properties of separators, the most interesting is Warburg's impedance (W), which characterizes diffuse processes inside the cell.

The main method of evaluating the data obtained by the impedance spectroscopy technique is the analysis of impedance hodographs, i.e., by studying the dependence of the imaginary parts of permittivity on the real parts of one set of Nyquist coordinates. Based on the data obtained from the hodograph, a cell's equivalent scheme can be constructed [25].

3. Results and Discussion

In the present work, two kinds of TEC architecture were investigated: thermocells with a salt bridge and thermocells with the coin cell CR2025 case. Both types have advantages and disadvantages, so the comparison was very informative.

Systems with a salt bridge are easy to maintain and can be precisely controlled under persistent large temperature differences. This provides an accurate measurement of the Seebeck coefficient. Systems in the coin cell CR2025 have two important application benefits: the assembly speed and the reproducibility of the results for cells in this type of case are much better than in other types of electrochemical systems.

Modification of Busofit-based electrode materials for both types of cells was carried out for two reasons. The first was for the formation of electrodes with a minimum contact resistance of the metal layer–Busofit. The second is a decrease in contact resistance between the Busofit layers, which determines the capacity of the electrolytic cell. In this case, the metal layer, in addition to low electrical resistance, must have a high porosity, which ensures the penetration of the electrolyte between the layers of Busofit and increases the total porosity. SEM images of the Busofit materials are shown in Figure 1.

Figure 1. SEM images of the Busofit materials: (**a**) Basic; (**b**) with Ag spraying; (**c**) with Ti spraying; (**d**) with infiltrated dispersion of TiO_2 nanoparticles.

Figure 1b,c show particles of deposited metal on the surface of the fiber. Figure 1d shows a thick coating. These particles can form sorption centers for redox reactions and have an influence on the interaction between electrolytes and the Busofit surface.

Plots of the maximum power density versus temperature difference between the hot and cold sides of the cell for the CR2025 case and cells with a salt bridge are shown in Figure 2.

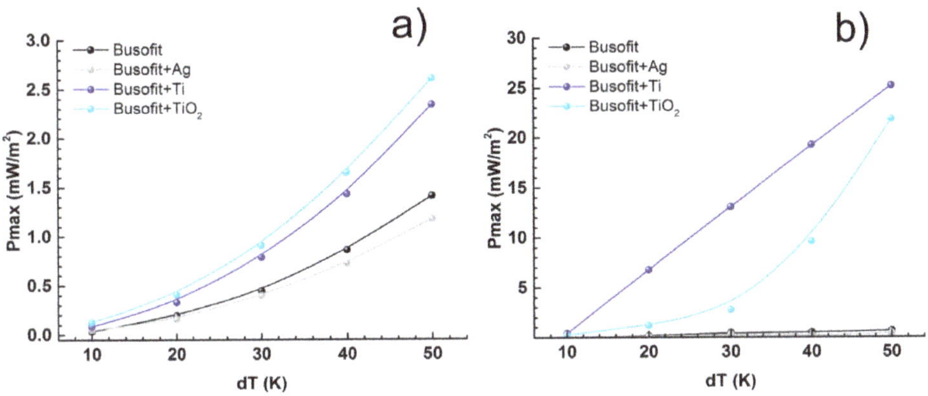

Figure 2. Plots of maximum power density versus temperature difference between the hot and cold sides of the cell (**a**) for cells with a salt bridge and (**b**) for cells in the coin cell CR2025.

The plots for the cells with a salt bridge (Figure 2a) show that the presence of the Ag modifier does not significantly affect the power of the system, while Ti deposition and TiO_2 dispersion significantly increase the output power (by about two times). This effect may be attributed to the reduction in the interface resistance between the electrodes and the electrolyte because of the catalytic effect of TiO_2 in the dispersion and the existence of oxide film on the surface of the deposited titanium.

In the coin cell case, the electrodes are very close to each other (about 1 mm apart); therefore, the actual temperature difference between the electrodes should be lower and the potential difference should also be lower. Nevertheless, the power output of the coin cell with the electrodes made of native Busofit and that of Busofit modified with Ag are close to the value obtained for the cells with a salt bridge. This can be explained by the lower resistance of the cell without a salt bridge and high exchange current values, due to the distance between the electrodes.

Based on the I–V plots of the thermal cells (Supplementary Materials Figures S1–S8), the values of the open circuit voltage and the short circuit current were compared (Table 1).

Table 1. The influence of surface modifier on the maximum values of U_{OC} and I_{SC} for different types of thermo-electrochemical cells (TECs) (temperature difference between the hot and cold sides is 50 K).

	Salt Bridge		**Coin Cell**	
	U_{OC}, mV	**I_{SC}, mkA**	**U_{OC}, mV**	**I_{SC}, mkA**
Native Busofit	71.5	12.0	11.1	51
Busofit with Ag	75.4	9.4	7.0	44
Busofit with Ti	75.4	17.7	11.3	1217
Busofit with TiO_2	72.6	21.5	19.0	682

The values of the open circuit voltage of cells with a salt bridge for all Busofit types are relatively close (Table 1). In this case, the voltage of the open circuit can be used to calculate the actual temperature difference of the electrodes using the formula: $\Delta T = U_{oc}/Se$, where Se is 1.4 mV/K.

For a coin cell, the real electrode temperature is significantly dependent on the thermal conductivity of the cell (electrodes, electrolyte, and current collector). Therefore, different Busofit modifications have different real temperature gradients, which are confirmed by U_{OC}. Hence, Busofit filled with TiO_2 which has lower thermal conductivity than native carbon fiber and carbon fiber with metallic silver or titanium showed higher UOC (19 mV) because of the higher real temperature gradient. The short circuit current is proportional to the equivalent serial resistance (ESR), which is obtained as the sum of the resistance of the electrolyte and the resistance to charge transfer through the electrode–electrolyte junction. It is obvious that the resistance of the electrolyte in coin cells is significantly lower than that of cells with a salt bridge, which is confirmed by the impedance data.

Nyquist plots (Figure 3a–e) have the shape of a semicircle and are equivalent to the parallel connection of capacitance and resistance [15]. In a TEC with an electrolyte containing Fe ions in the heterovalent state, the reaction proceeds on the surface of the hot electrode,

$$Fe(CN)_6{}^{3-} - e^- \rightarrow Fe(CN)_6{}^{4-} \tag{2}$$

and on the surface of the cold electrode,

$$Fe(CN)_6{}^{4-} + e^- \rightarrow Fe(CN)_6{}^{3-} \tag{3}$$

The impedance of a TEC consists of R_s—electrolyte resistance; $CPE_{dl(c)}$, $CPE_{dl(h)}$—constant phase elements for cold and hot electrodes that correspond to capacities of an electrical double layer at the electrode–electrolyte interface. $R_{ct(c)}$ and $R_{ct(h)}$ are the charge transfer resistance of cold electrode–electrolyte and hot electrode–electrolyte junctions, respectively.

The complete impedance circuit (Figure 3e) can be simplified provided that the resistances and capacitances of the electric double layer of the hot and cold electrodes differ by small values. During reactions (2) and (3), the potential of the electrodes for carbon materials changes. These changes are accompanied by the accumulation of an pseudo-intensity of redox reactions on the surface of the Busofit material:

$$> C\text{-}OH \leftrightarrow C{=}O + H^+ + e^-, \text{-}COOH \leftrightarrow \text{-}COO + H^+ + e^- \text{ and } > C{=}O + e^- \leftrightarrow > C\text{-}O\text{-} \tag{4}$$

At a temperature gradient, the TEC gives off charge as a supercapacitor, i.e., the current is inversely proportional to the ESR, which is the sum of the resistance of the electrolyte and the resistance of the charge transfer of the electrodes. To confirm the hypothesis of limiting resistance and to calculate the thermoelectric conversion efficiency indicator, an analysis of the results by impedance spectroscopy was carried out using the method of determination parameters of the equivalent circuit. The equivalent scheme of the TEC and dependences of the cell's resistance are shown in Figure 3.

On basis of these results, the dependence of the resistance of the electrolyte on the temperature difference of the TEC was established.

The calculated values of the elements of the equivalent scheme (Figure 3a) are shown in Table 2.

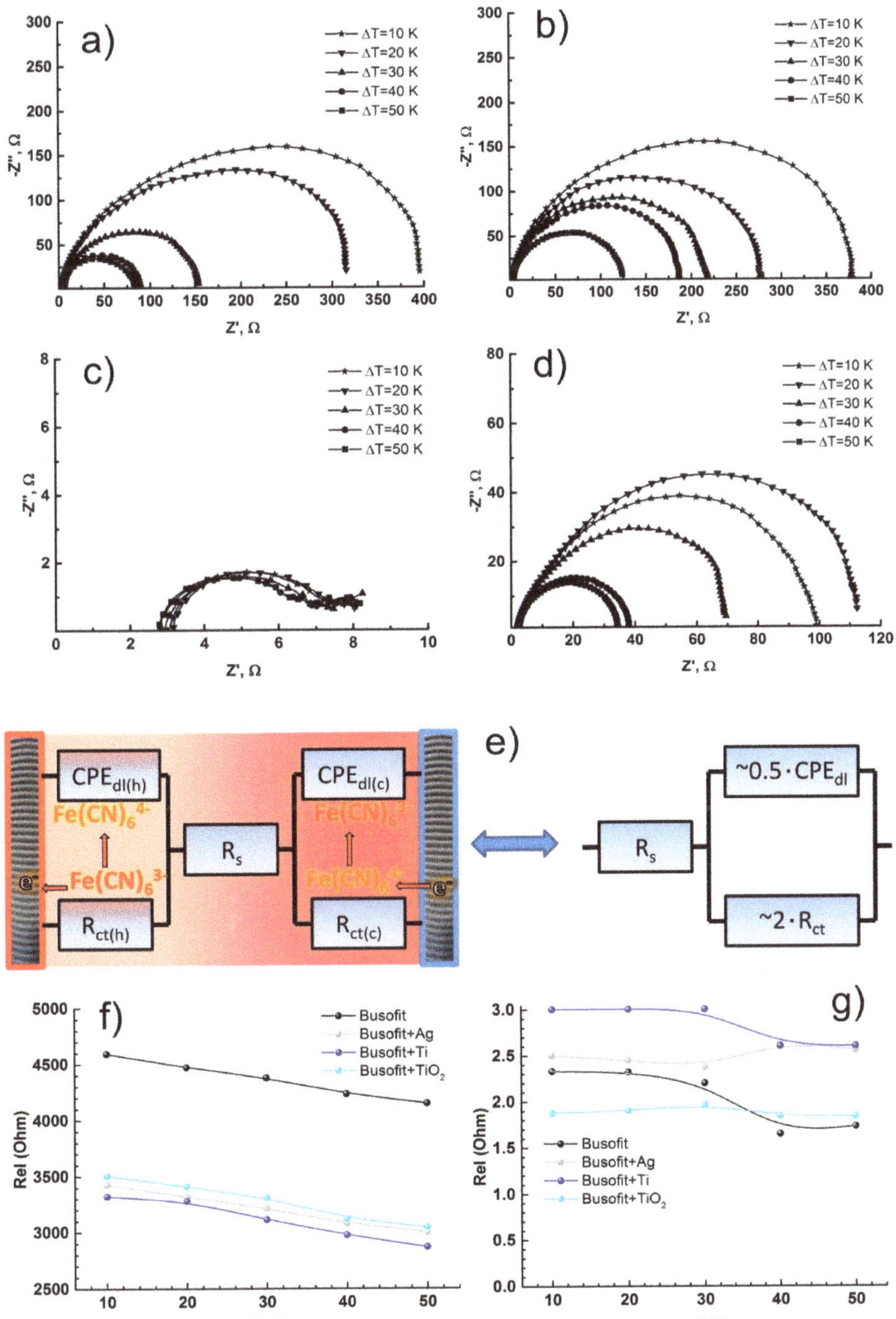

Figure 3. Nyquist plots for cells in a coin cell CR2025: (**a**) Busofit, (**b**) Busofit Ag, (**c**) Busofit Ti, and (**d**) Busofit TiO_2; (**e**) equivalent scheme of a TEC (**e**); (**f**) the dependence of the resistance of the electrolyte on the temperature difference between the electrodes for cells with a salt bridge and (**g**) for cells in a coin cell CR2025.

Table 2. Elements of the equivalent scheme of the studied TECs.

Type of Electrodes	R_s (Ω)	C_{dl} (mkF)	R_{ct} (Ω)
Busofit (Salt bridge)	4150	6	1100
Busofit (Coin cell)	2	3	80
Busofit+Ag (Salt bridge)	3000	3	4090
Busofit+Ag (Coin cell)	2	3	120
Busofit+Ti (Salt bridge)	2870	3	201
Busofit+Ti (Coin cell)	2	10	5
Busofit+TiO_2 (Salt bridge)	3040	6	430
Busofit+TiO_2 (Coin cell)	2	3	30

From these results, it is seen that the resistance of the electrolyte in systems with a salt bridge is thrice that in a system in a coin cell CR2025. This explains the difference in the values of specific power and current density between systems with the same electrodes but assembled in different cell architectures.

A comparison of the resistance of the electrode process for various modifications of the Busofit fiber shows that the resistance of the fiber modified with titanium is 15 to 20 times lower than the resistance of the base fiber or the resistance of the one modified with silver. At the same time, fiber modified with titanium oxide powder has a higher resistance, but it is significantly lower than the starting material.

Such significant differences in resistance are reflected in parameters such as short circuit current and output power. On the one hand, the short circuit current at the maximum temperature gradient for cells in the coin cell was, on average, twice as high as that for cells with a salt bridge. In addition, the short circuit current of Busofit fibers modified with titanium (Ti and TiO_2) was about 20–30 times higher than that for base fibers or fibers with silver.

Busofit contact resistance consists of the internal resistance of individual fibers, the total resistance of each contact between the fibers, and the charge transfer resistance across the electrode–electrolyte junction. It is obvious that Busofit is in a compressed state, i.e., assembled in a coin-cell form, and has the lowest total resistance of all contacts between the fibers compared with the circuit with a bridge, as confirmed by the data in Table 1. In this case, the silver coating (sputtering) on the surface of Busofit increases the impedance of the contacts and is apparently associated with a rapid oxidation of silver and the formation of oxide with high resistance. In contrast, titanium sputtering significantly reduced contact resistance from 80 to 5 Ω (Table 2).

On the other hand, the open circuit voltage (Supplementary Materials) for cells in the coin cell is lower, on average, by 5–15 times compared to cells with a salt bridge, which is associated with a lower temperature gradient.

Thus, regardless of the type of cell architecture, modification of the electrode material with titanium or titanium oxide significantly improves the short circuit current of any type cell. This effect is associated with the catalytic effect of titanium oxide on the oxidation reaction occurring on the hot electrode.

To confirm the hypothesis of the catalytic effect of titanium dioxide on the surface of the electrode material Busofit, the resistance value of the electrode process R_{ct} was analyzed. The dependence of the R_{ct} value on temperature at the Arrhenius coordinates for the Busofit-based fiber systems with sputtered titanium, silver, and infiltrated titanium dioxide in the cells of the coin cell is shown in Figure 4a. The dependence of R_{ct} on the absolute temperature at the Arrhenius coordinates is linear. The activation energy of the electrode process is depicted in the summary plot (Figure 4b). The activation energy of the electrode process in system with titanium dioxide is lower than in other systems.

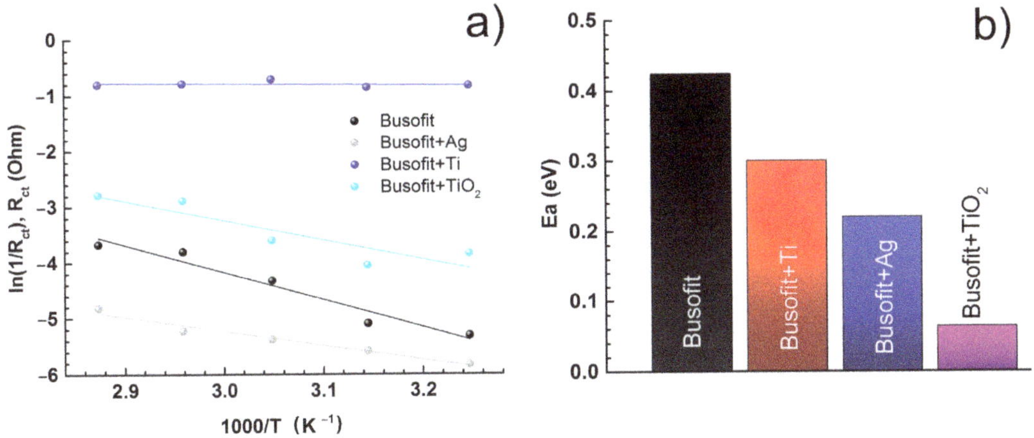

Figure 4. Dependence of the resistance of the electrode process on the temperature at the (**a**) Arrhenius coordinates, and (**b**) values of the activation energy of the electrode process in systems with the base fabric Busofit with Ag coating, Ti coating, and infiltrated TiO_2.

The efficiency of thermoelectric conversion was calculated by the method in [26]. The dependence of the equivalent parameter ZT on the temperature gradient of the electrodes is shown in Figure 5. Thus, ZT varies with change in resistance.

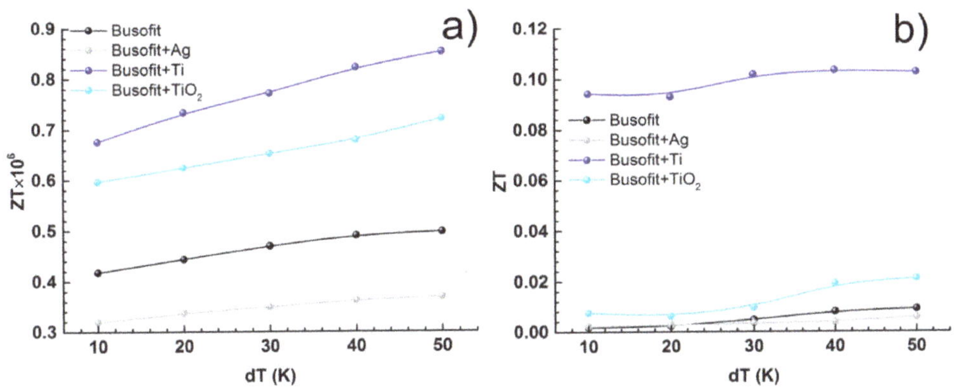

Figure 5. The dependence of the equivalent parameter ZT on the temperature gradient (**a**) for cells with a salt bridge and (**b**) for cells in a coin cell CR2025.

In [27], it was shown that the values of ZT typical for solid-state inorganic thermal generators are 1 or less at temperatures of about 800 °C. For example, for Bi_2Ti_3, ZT is in the range of 0.5–0.8; for composites based on rare-earth metals, ZT is in the range of 0.75–1.2, and for polymeric materials, ZT is in the range of 0.2–0.42.

Based on the ZT values, the dependence of the thermoelectric conversion efficiency was established (Figure 6). Plots of the thermoelectric conversion efficiency reveal that the efficiency of the system with a salt bridge is very low. This can be explained by the high resistance of the electrolyte in the connecting tubes. The system with carbon fiber Busofit coated with titanium in a coin cell CR2025 showed the highest efficiency of 1.37%, with a temperature difference of 50 K. In a previously published review [4], values for typical thermo-electrochemical cells <1% were shown and it was noted that an efficiency of 2–5% will be sufficient for applied use.

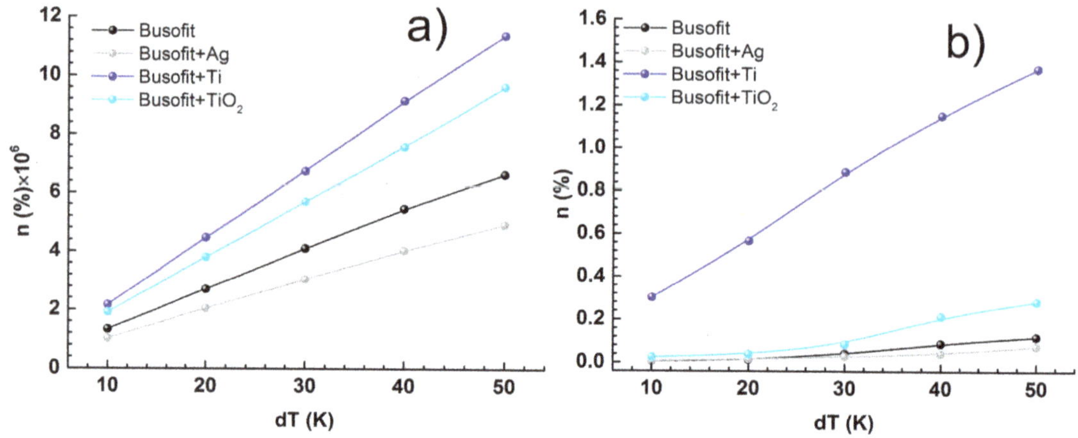

Figure 6. The dependence of the efficiency of the studied cells on the temperature gradient of the electrodes (**a**) for cells with a salt bridge and (**b**) for cells in a coin cell CR2025.

4. Conclusions

In this work, the properties of TECs with two different cell architectures (electrochemical cell with a salt bridge and a coin cell case) based on modified carbon fiber electrodes were studied. The electrochemical impedance of TECs for various temperature gradients was described by an equivalent scheme; the limiting impedance components for the salt bridge and coin cell systems were calculated and explained for three parameters (R_s, $CPE_{dl(c)}$, $CPE_{dl(h)}$, $R_{ct(c)}$ and $R_{ct(h)}$, look at the Figure 3e).

Close distance between electrodes causes low values of electrolyte resistance in the coin cell. At the same time, different thermal conductivity of carbon fibers modified with Ag, Ti, and TiO_2 causes different real temperature in the electrode–electrode contact zone and significantly decreases the real temperature gradient. This effect is well observed from the dependence of the open circuit voltage (U_{OC}) on the type of modifier in the cell in Table 1, which is significantly higher for cells with a salt bridge. Nevertheless, high resistance of the salt bridge decreases the exchange current, and for this reason, the relative efficiency and power output of the coin cell system exceed the values of these parameters in cells with a salt bridge.

The catalytic effect of the Busofit modifiers was confirmed by a change in the activation energies, which were calculated from the Arrhenius dependences for contact resistance. The TiO_2 Busofit modifier, with an activation energy about 0.3 eV, has the greatest catalytic effect. The highest value of specific electric power of the cells was achieved in the coin cell system with Busofit+Ti electrodes. This can be caused by the combination of the high conductivity of titanium metal, which reduces the resistance between individual carbon filaments, and the catalytic effect of a thin oxide film on its surface. The values of the specific power and the efficiency were 25.2 mW/m^2 and 1.37%, respectively.

Supplementary Materials: The following are available online at https://www.mdpi.com/2071-1050/13/3/1377/s1, Figure S1: (a) I-V curves and (b) P-V curves for cells with salt bridge and basic.fiber Busofit respectively; Figure S2: (a) I-V curves and (b) P-V curves for cells with salt bridge and Ag-sputtering Busofit respectively; Figure S3: (a) I-V curves and (b) P-V curves for cells with salt bridge and Ti-sputtering Busofit re-spectively; Figure S4: (a) I-V curves and (b) P-V curves for cells with salt bridge and Busofit with infiltrated dispersion of TiO_2 nanoparticles respectively; Figure S5: (a) I-V curves and (b) P-V curves for cells in CR 2025 case and basic fiber Busofit re-spectively; Figure S6: (a) I-V curves and (b) P-V curves for cells in CR 2025 case and Ag-sputtering Busofit respectively; Figure S7: (a) I-V curves and (b) P-V curves for cells in CR 2025 case and Ti-sputtering

Busofit re-spectively; Figure S8: (a) I-V curves and (b) P-V curves for cells in CR 2025 case and Busofit with infiltrated dispersion of TiO_2 nanoparticles respectively.

Author Contributions: Conceptualization and methodology, N.G.; validation, N.K. (Natalya Kovyneva) and M.V.; formal analysis, E.B.; software and visualization, D.A.; investigation, O.G.; resources, A.G.; data curation and writing—original draft preparation, N.K. (Nikolay Kiselev); writing—review and editing, P.O.; supervision and project administration, I.B. All authors have read and agreed to the published version of the manuscript.

Funding: This research received no external funding.

Institutional Review Board Statement: Not applicable.

Informed Consent Statement: Not applicable.

Data Availability Statement: Not applicable.

Conflicts of Interest: The authors declare no conflict of interest.

References

1. Im, H.; Kim, T.; Song, H.; Choi, J.; Park, J.S.; Ovalle-Robles, R.; Yang, H.D.; Kihm, K.D.; Baughman, R.H.; Lee, H.H.; et al. High-efficiency electrochemical thermal energy harvester using carbon nanotube aerogel sheet electrodes. *Nat. Commun.* **2016**, *7*, 10600. [CrossRef]
2. Dupont, M.F.; Macfarlane, D.R.; Pringle, J. Thermo-electrochemical cells for waste heat harvesting—Progress and perspectives. *Chem. Commun.* **2017**, *53*, 6288–6302. [CrossRef]
3. Burmistrov, I.; Kovyneva, N.; Gorshkov, N.; Gorokhovsky, A.; Durakov, A.; Artyukhov, D.; Kiselev, N. Development of new electrode materials for thermo-electrochemical cells for waste heat harvesting. *Renew. Energy Focus* **2019**, *29*, 42–48. [CrossRef]
4. Zhang, L.; Kim, T.; Li, N.; Kang, T.J.; Chen, J.; Pringle, J.M.; Zhang, M.; Kazim, A.H.; Fang, S.; Haines, C.; et al. High Power Density Electrochemical Thermocells for Inexpensively Harvesting Low-Grade Thermal Energy. *Adv. Mater.* **2017**, *29*, 1605652. [CrossRef]
5. Romano, M.S.; Li, N.; Antiohos, D.; Razal, J.M.; Nattestad, A.; Beirne, S.; Fang, S.; Chen, Y.; Jalili, R.; Wallace, G.G.; et al. Carbon Nanotube—Reduced Graphene Oxide Composites for Thermal Energy Harvesting Applications. *Adv. Mater.* **2013**, *25*, 6602–6606. [CrossRef]
6. Zhou, Y.; Qian, W.; Huang, W.; Liu, B.; Lin, H.; Dong, C. Carbon Nanotube-Graphene Hybrid Electrodes with Enhanced Thermo-Electrochemical Cell Properties. *Nanomaterials* **2019**, *9*, 1450. [CrossRef]
7. Hirai, T.; Shindo, K.; Ogata, T. Charge and Discharge Characteristics of Thermochargeable Galvanic Cells with an $[Fe(CN)_6]^{4-}/[Fe(CN)_6]^{3-}$ Redox Couple. *J. Electrochem. Soc.* **1996**, *143*, 1305. [CrossRef]
8. Artyukhov, D.; Kiselev, N.; Gorshkov, N.; Burmistrov, I. Research of the influence of electrolyte concentration on ther-mo-electrochemical cells efficiency. *Proc. Environ. Sci. Eng. Manag.* **2019**, *6*, 319–327.
9. Duan, J.; Feng, G.; Yu, B.; Li, J.; Chen, M.; Yang, P.; Feng, J.; Liu, K.; Zhou, J. Aqueous thermogalvanic cells with a high Seebeck coefficient for low-grade heat harvest. *Nat. Commun.* **2018**, *9*, 1–8. [CrossRef]
10. Hu, R.; Cola, B.A.; Haram, N.; Barisci, J.N.; Lee, S.; Stoughton, S.; Wallace, G.; Too, C.; Thomas, M.; Gestos, A.; et al. Harvesting Waste Thermal Energy Using a Carbon-Nanotube-Based Thermo-Electrochemical Cell. *Nano Lett.* **2010**, *10*, 838–846. [CrossRef]
11. Salazar, P.F.; Kumar, S.; Cola, B.A. Design and optimization of thermo-electrochemical cells. *J. Appl. Electrochem.* **2013**, *44*, 325–336. [CrossRef]
12. Wu, J.; Black, J.J.; Aldous, L. Thermoelectrochemistry using conventional and novel gelled electrolytes in heat-to-current thermocells. *Electrochim. Acta* **2017**, *225*, 482–492. [CrossRef]
13. Zhou, H.; Liu, P. High Seebeck Coefficient Electrochemical Thermocells for Efficient Waste Heat Recovery. *ACS Appl. Energy Mater.* **2018**, *1*, 1424–1428. [CrossRef]
14. Burmistrov, I.; Gorshkov, N.; Kovyneva, N.; Kolesnikov, E.; Khaidarov, B.; Karunakaran, G.; Cho, E.-B.; Kiselev, N.; Artyukhov, D.; Kuznetsov, D.; et al. High seebeck coefficient thermo-electrochemical cell using nickel hollow microspheres electrodes. *Renew. Energy* **2020**, *157*, 1–8. [CrossRef]
15. Burmistrov, I.; Gorshkov, N.; Kiselev, N.; Artyukhov, D.; Kolesnikov, E.; Khaidarov, B.; Yudni, A.; Karunakaran, G.; Cho, E.-B.; Kuznetsov, D.; et al. Data on the current-voltage dependents of nickel hollow microspheres based thermo-electrochemical in alkaline electrolyte. *Data Brief* **2020**, *31*, 105770. [CrossRef]
16. Gunawan, A.; Lin, C.-H.; Buttry, D.A.; Mujica, V.; Taylor, R.A.; Prasher, R.; Phelan, P. Liquid Thermoelectrics: Review of Recent and Limited New Data of Thermogalvanic Cell Experiments. *Nanoscale Microscale Thermophys. Eng.* **2013**, *17*, 304–323. [CrossRef]
17. Benji, M.; Alam, K. Carbon nanotubes and nanofibers in composite materials. *Sampe J.* **2002**, *38*, 59–70.
18. Miyasaka, K.; Watanabe, K.; Jojima, E.; Aida, H.; Sumita, M.; Ishikawa, K. Electrical conductivity of carbon-polymer composites as a function of carbon content. *J. Mater. Sci.* **1982**, *17*, 1610–1616. [CrossRef]
19. Feng, L.; Xie, N.; Zhong, J. Carbon Nanofibers and Their Composites: A Review of Synthesizing, Properties and Applications. *Materials* **2014**, *7*, 3919–3945. [CrossRef]

20. Salazar, P.F.; Kumar, S.; Cola, B.A. Nitrogen- and Boron-Doped Carbon Nanotube Electrodes in a Thermo-Electrochemical Cell. *J. Electrochem. Soc.* **2012**, *159*, B483–B488. [CrossRef]
21. Alzahrani, H.A.; Buckingham, M.A.; Marken, F.; Aldous, L. Success and failure in the incorporation of gold nanoparticles inside ferri/ferrocyanide thermogalvanic cells. *Electrochem. Commun.* **2019**, *102*, 41–45. [CrossRef]
22. Gong, K.; Du, F.; Xia, Z.; Durstock, M.; Dai, L. Nitrogen-Doped Carbon Nanotube Arrays with High Electrocatalytic Activity for Oxygen Reduction. *Science* **2009**, *323*, 760–764. [CrossRef]
23. Bae, K.M.; Yang, H.D.; Tufa, L.T.; Kang, T.J. Thermobattery based on CNT coated carbon textile and thermoelectric electrolyte. *Int. J. Precis. Eng. Manuf.* **2015**, *16*, 1245–1250. [CrossRef]
24. Im, H.; Moon, H.G.; Lee, J.S.; Chung, I.Y.; Kang, T.J.; Kim, Y.H. Flexible thermocells for utilization of body heat. *Nano Res.* **2014**, *7*, 443–452. [CrossRef]
25. Macdonald, J.R.; Barsoukov, E. *Impedance Spectroscopy: Theory; Experiment; and Applications*, 2nd ed.; Wiley & Sons: New York, NY, USA, 2005; pp. 289–305.
26. Goldsmid, H.J. *Introduction to Thermoelectricity*; Springer: Berlin, Germany, 2010; p. 46.
27. Champier, D. Thermoelectric generators: A review of applications. *Energy Convers. Manag.* **2017**, *140*, 167–181. [CrossRef]

sustainability

MDPI

Article

Toward Sustainable Environmental Management of Healthcare Waste: A Holistic Perspective

Nouf Sahal Alharbi [1,*], Jawaher Haji Alhaji [2] and Malak Yahia Qattan [2]

[1] Department of Health Administration, College of Business Administration, King Saud University, KSA, ZIP 4545, Riyadh 11451, Saudi Arabia
[2] Department of Health Sciences, College of Applied Studies and Community Service, King Saud University, KSA, ZIP 4545, Riyadh 11451, Saudi Arabia; jalhejjy@ksu.edu.sa (J.H.A.); mqattan@ksu.edu.sa (M.Y.Q.)
* Correspondence: noufsahal@ksu.edu.sa; Tel.: +96-6114693997

Abstract: The management of healthcare waste requires a sustained and holistic approach involving a range of parties. This is challenging for governments, especially in developing countries, where waste management systems have limited capacities for addressing the issue. Using Saudi Arabia as a case study, this paper followed a multi-method approach, including policy analysis, observation, semi-structured interviews, and a focus group, to explore the country's healthcare waste management system. The study estimated that Saudi government hospitals across the country, every year, throw away in landfills paper (27,000 tons), plastic (15,000 tons), food (10,000 tons), glass (8000 tons), and metal (7000 tons). Regrettably, all these tons of materials end up in landfills without any form of recycling. A number of challenges were identified, reflecting mainly the lack of a legal framework, waste training, coordination among stakeholders, and social responsibility. This study generated new knowledge about waste management systems by exploring how their performance is shaped by the processes occurring at the policy, organization, and individual levels.

Keywords: sustainability; healthcare waste management; Saudi Arabia; social responsibility; assessment method; policy analysis

Citation: Alharbi, N.S.; Alhaji, J.H.; Qattan, M.Y. Toward Sustainable Environmental Management of Healthcare Waste: A Holistic Perspective. *Sustainability* **2021**, *13*, 5280. https://doi.org/10.3390/su13095280

Academic Editors: Rita Khanna, Yury Konyukhov, Igor Burmistrov and Chunjiang An

Received: 3 March 2021
Accepted: 6 May 2021
Published: 9 May 2021

1. Introduction

Healthcare service facilities contribute to the control, prevention, and treatment of diseases and so represent an important dimension of sustainability. The health sector in any country consumes a significant part of financial and human resources, thus reflecting the economic dimension of sustainability. Consequently, healthcare organizations consume a significant amount of material and inevitably create waste, which affects environmental sustainability [1,2]. Addressing healthcare waste not only significantly creates a considerable number of benefits to the environment but also contributes to cost saving and improves the public image of the parties involved [3,4].

Healthcare wastes usually result from the provision of medical care and its associated support services, such as nutrition and maintenance. According to the World Health Organization (WHO), around 15% of healthcare waste is estimated to be hazardous medical waste (HMW), such as infectious, radioactive, or pathological materials, while the vast majority of waste is considered municipal solid waste (MSW), such as that coming from households in the form of food, paper, and plastic [5]. Internationally, the generation of healthcare waste has increased significantly due to population growth and the concomitant increase in the number of healthcare facilities [6]. Such development has created an environmental and economic burden on countries around the world. According to Zimmer and McKinley, healthcare waste is the fourth-largest contributor of mercury in the environment, and managing MSW and HMW costs USD 0.05 and USD 1.25 a pound, respectively [7].

Due to the economic and environmental impacts of waste, early debates on the health industry and its sustainability have mostly focused on waste elimination and recycling as critical strategies for health systems that promote environmentally friendly practices. This is the main objective of the United Nations Sustainable Development Goals (UNSDGs), whose aim is to ensure a sustainable future for all by creating a balance between social, economic, and environmental development and improving system performance by meeting current and future stakeholders needs [8,9]. The growth of environmental awareness and the development of robust regulation, coupled with the current need to reduce public spending, have all accentuated environmental issues for the health sector. Therefore, developing and implementing sustainable healthcare waste management requires a holistic approach involving a range of parties, including policymakers, government and non-government organizations, and the community [2,10–12].

With regard to healthcare management within healthcare organizations, several studies have highlighted environmental measures for resource use minimization and waste reduction. However, few studies have used a holistic approach that includes the political, organizational, and individual attempts toward sustainable waste management. This paper focuses on the Saudi Arabia Ministry of Health's (MOH) waste management system as a case study, with the main objective of reviewing the status of the UN Sustainable Development Goals and the country's alignment with its Vision 2030. The Saudi government is aiming to achieve environmental sustainability by preserving natural resources and increasing the efficiency of waste management [13]. In this respect, the country provides an excellent research context to explore the extent to which the sustainability objectives are being translated into practice. Therefore, this paper set out (1) to explore the key factors that determine healthcare waste sustainability by applying a holistic methodology and (2) to suggest recommendations for improving the waste management system.

2. Materials and Methods

Saudi Arabia is the largest country in the Arabian Gulf as well as the Middle East, with a territory of 2,149,690 km^2. The country is a member of the Gulf Cooperation Council (GCC) along with Oman, Bahrain, the United Arab Emirates, Qatar, and Kuwait. The capital of Saudi Arabia is Riyadh, where this study was conducted [14]. In Saudi Arabia, the MOH is considered the main healthcare provider, with 284 hospitals and 2390 primary healthcare (PHC) clinics [15]. Riyadh was selected for this study as it is the business center of the country where all ministries and main waste management institutions, as well as 20% of the total healthcare facilities, are situated. According to the latest statistics, there are 17 government hospitals, 37 private hospitals, 125 government primary care centers, and 818 private clinics in Riyadh [16].

To assess the sustainability of waste production and consumption, it is best to follow a multidisciplinary approach to collect information so as to incorporate different data perspectives [17]. Therefore, this study used a combination of four different research methods in order to gather data and explore the opportunities and challenges in healthcare waste management:

- Analyzing all existing official documents about healthcare waste policies and procedures in Saudi Arabia;
- Undertaking a rapid assessment observation of healthcare waste management policies and practices;
- Conducting semi-structured interviews of key informants in health waste management;
- Convening a focus group with healthcare practitioners.

2.1. *Document Analysis of National Policy, Regulations, and Other Information*

The macro level of health waste management in Saudi Arabia was first assessed by analyzing relevant existing documents, including guidelines and laws concerning health waste management and background information about health waste records and statistics. Given that the published policy documents were not indexed or listed in any academic databases in a standardized and timely manner, the data search began with an examination of the gray literature: government and mass media websites. To increase the number of documents and ensure reliability, we expanded the scope of the search for information to the GCC Health Council and the WHO, as these organizations are the most globally recognized ones that report information regarding healthcare waste. In addition, a literature search for academic papers on healthcare waste management in Saudi Arabia was conducted on Web of Science databases using the following keywords: waste, hazardous, healthcare waste, hospital waste, medical waste, clinical waste, environmental management, legislation, policies, KSA, Saudi Arabia, and GCC countries. These were separate from the official policy documents about HCW. National and local reports, guidelines, and national conference reports were also examined.

Most of the documents were collected from official governmental websites; however, some statistics were not easily identified through online data searches and were collected through interviews of governmental officers. Following the principles of the framework synthesis approach [18], we deductively extracted, summarized, and synthesized data obtained from the documents according to the corporate environmental performance framework that conceptualizes sustainable performance into five environmental management performance elements (inputs: policy, objectives, processes, organizational structure, and environmental monitoring) and environmental operational performance (outcomes). Policy constitutes a macro-level commitment and responsibility toward the improvement of environmental operational performance. Objectives refer to specific environmental goals and targets in order to translate environmental policy into action. Processes refer to actions designed to improve environmental operational performance. The organizational structure refers to the formal organizational structure implemented to achieve environmental goals. Environmental monitoring refers to corrective actions and procedures that ensure continuous improvement of environmental operational performance. [19]. The focus was only on waste as one of the framework's sustainable environmental performance indicators. This framework was considered a useful one for the study, hence its selection.

2.2. Rapid Assessment of Healthcare Waste (Observation)

Observation of the health waste management process was assessed by adapting and modifying the healthcare waste management tool for rapid assessment [20]. The checklist includes 6 criteria of waste management practice and 21 indicators. The tool ranks the performance of each criterion by the level of sustainability ranging from 0 (operating in a totally unsustainable manner with reluctance to change) to 4 (operating in a way that displays all the characteristics normally associated with sustainable development) (Tables 1 and 2).

2.3. Semi-Structured Interview of Key Informants

To identify the challenges and opportunities in improving the sustainability of healthcare waste management, semi-structured interviews were conducted with some of the main parties involved in healthcare waste management in Saudi Arabia. The main stakeholders were identified by reviewing official documents in the first phase of the study, and then their number was supplemented by new participants by applying the snowballing technique [21]. There were finally a total of 12 interviewees: 3 from the MOH, 2 from private waste management companies, 3 environmentalists working with public authorities, and 4 academics working in the field of healthcare waste management.

2.4. Focus Group with Healthcare Providers

A focus group session was conducted in one of the largest hospitals in Riyadh, in which 10 healthcare workers, considered critical parties in healthcare waste management, were brought together as a group. They included physicians, nurses, laboratory specialists, dieticians, and a receptionist. Half of them were males, and the age ranged between 26 and 51 years. To ensure that participants could speak openly about the issues, persons who worked in the environmental department were excluded. The session lasted for 40 min and was conducted in a conversational style. The main questions were developed based on the data gathered during the documentary analysis and interviews. The discussion covered waste minimization, rules and regulations, and the training received on waste disposal. All participants were assured that their responses would remain confidential and would have no influence on their work evaluation. The focus group was convened, audiotaped, and transcribed by the researcher with the name of the participants and the hospital coded for confidentiality.

Table 1. Healthcare waste rapid assessment guidelines.

Waste Management Criteria	Waste Management Indicators
General management strategy	Hospital waste management policy or strategy Special budget for waste management Operative staff for waste management Training on waste management
Waste collection and segregation	Number of containers Type of containers Color coding Waste collection and segregation Quantity of waste
Waste storage	Washing and disinfecting of storage containers Separate packaging and transport of waste Waste storage in areas with built-in roofs
Waste recycling	What is recycled? What is reused?
Waste treatment	Waste treatment in a purpose-built waste treatment plant Autoclaving of lab waste Chemical disinfection of body fluids Incineration outside Encapsulation of sharps
Off-site disposal	Transport in enclosed vehicles Final destination

Table 2. Sustainable operational performance levels.

Level 0	Operating in a Totally Unsustainable Manner with Reluctance to Change
Level 1	Generally operating in an unsustainable manner, although some evidence of awareness and a willingness to change
Level 2	Operating in a manner with some aspects that are considered sustainable and others that are considered unsustainable
Level 3	Generally operating in accordance with sustainable development, but some aspects are not ideal
Level 4	Operating in a way that displays all the characteristics normally associated with sustainable development

2.5. Data Triangulation

This study adopted a multi-method approach, where the data were collected and analyzed independently. Therefore, triangulation was used to integrate overlapping of the multi-method findings, according to Brewer and Hunter [22] (Figure 1). For example, the focus group provided a snapshot of the healthcare workers' attitude toward waste segregation, document analysis and observation showed the outcome of their behavior, and the interviews gave a broader view of the strategies that could be adopted by the policymakers. This study synthesized the findings according to the health service levels: macro (policy), meso (healthcare organization), and micro (individuals).

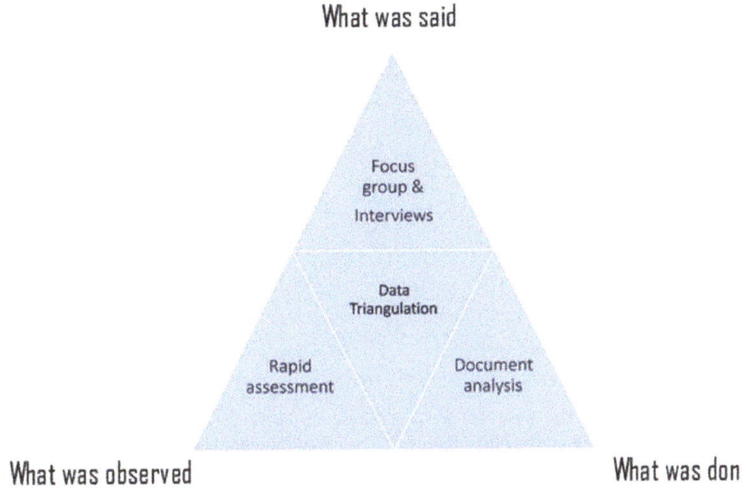

Figure 1. An illustration of the study data triangulation methodology.

3. Results

3.1. Macro Level: Analysis of Regulation, Guidelines, and National Data

The definition of the healthcare waste policy, regulations, and standard system began about two decades ago. In 1998, the Saudi MOH passed regulations defining HMW handling, treatment, and disposal. One year later, the GCC endorsed a unified HMW regulation among all Gulf countries, including Saudi Arabia. Since the waste management national policy was developed, it has been revised twice, in 2005 and 2019 [23–25]. During the interviews, the participants mentioned four main policymaking parties: the MOH, the Ministry of Environmental Water and Agriculture, the Ministry of Municipal and Rural Affairs, and private waste management companies. All participants agreed that there was a need for a legal framework to enhance the level of coordination among these parties. An environmentalist in an academic job said:

"The Ministry of Environment has specific legal requirements about medical waste, but due to the lack of coordination between it and the other parties, it is not sufficiently monitored."

The Saudi government recognized the need for a regulatory body to enhance the multisectoral collaboration between all the waste management parties, including those involved in healthcare waste in the country. Therefore, the Saudi Waste Management Center (SWMC) was established in May 2019 to organize and supervise waste management regulations and practices. In this organization, the representatives on the center's board came from different ministries, namely those for the environment, rural affairs, economy, health, and energy, as well as a representative from the general investment's authority [26]. The national healthcare waste plans and initiatives are presented in a time line in Figure 2.

To explore the sustainability of healthcare waste management at the macro level, we adopted a modified version of the elements of the organizational environmental performance framework [19]. With regard to the first framework input element (policy), the national policy classified HMW into eight groups: infectious, pathological, sharp, pharmaceutical, chemical, genotoxic, radioactive, and pressurized containers. The existing HMW policy also covered waste collection, storage, transportation, treatment, and disposal practices. However, the classification and management of MSW were totally absent from the policy statements. In addition, the policy did not state any sustainable targets or objectives in terms of outcome measures that should be achieved by healthcare facilities [25]. With regard to the third element (process), document analysis showed that Saudi healthcare waste is affected by practical obstacles such as a lack of awareness on the part of health workers of the HMW color code, segregation, disinfection, and storage. This was evident both in hospitals [27–29] and among PHC workers [30,31].

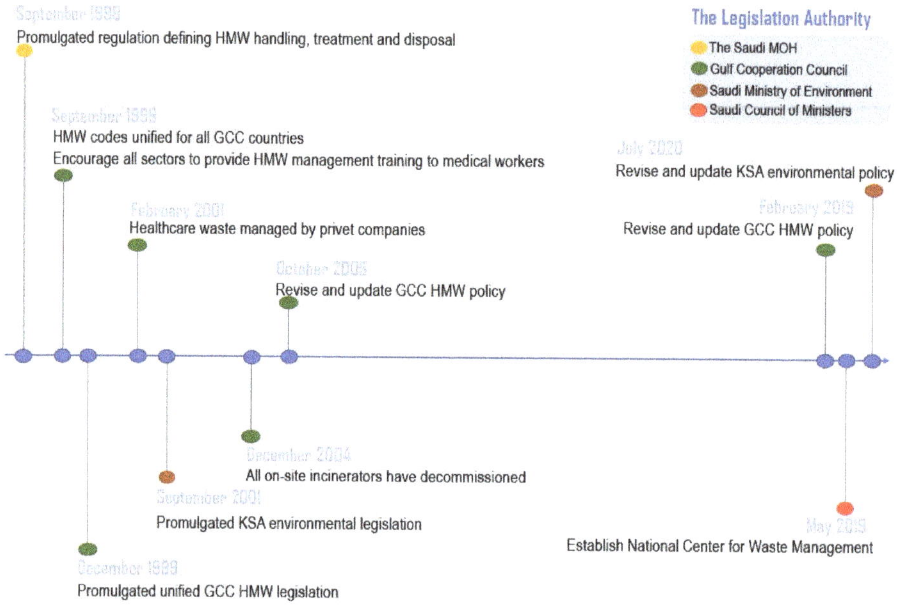

Figure 2. Time line of the national healthcare waste plans and initiatives.

With regard to organizational structure, document analysis and focus group results revealed that to achieve sustainable management of healthcare waste, there is a need to strengthen the role of the health worker. This could be done by the workers improving their sustainable environmental practices and them being made aware, through training, of the environmental impact of their work. Furthermore, multisector collaboration between the MOH and other parties in Saudi Arabia should be considered [27,29,31,32]. During the focus group, the health workers suggested that training on waste handling should be a requirement for renewing their professional licenses. According to one health worker:

"I had medical waste training in the first year of my work in the hospital. I think if there was regular training on general waste sorting, like paper and plastic, and if we made this training a requirement to renew the professional license, we will increase the level of waste sustainability in the hospital."

Auditing (or environmental monitoring) is the final key input element of the environmental sustainability framework. Both document analysis and direct observations revealed the absence of waste audit practices, inspections, waste weighing, and a continuous feedback and communication channel between the different parties [29,32].

All the above-mentioned elements are responsible for environmental sustainability and enhancement of the operational performance of the environment (Figure 3). To obtain a comprehensive picture of the production of healthcare waste on a national level, the latest data were extracted from monitoring records and mapped with the healthcare facilities and services in each region of the country (Appendix A). The official statistical data report was from the MOH, and the value of the data covers the period from January to December 2018. The data were obtained from the environmental health department where waste was segregated daily based on its source and weighed before disposal at the landfill site. The data reflected the weight of the HMW taken from all the hospitals and PHC clinics around the country. The total estimated HMW produced by hospitals and PHC clinics were 21,016 and 1347 tons per year, respectively. However, no weighing or reliable estimations were available for the MSW produced. Giving the fact that only 15 to 20% of the total healthcare waste is hazardous [5], the MSW estimated for all hospitals and PHC clinics, based on 80% of the available data, was 84,064 and 5390 tons per year, respectively (Table 3). To enable a global comparison, we also estimated the waste per bed and per visit. Given the fact that the bed occupancy rate in Saudi Arabia was 77% [15], according to the secondary data provided by the MOH, the estimated HMW and MSW for hospitals were 1.7 and 6.8 kg/bed/day, respectively, and those for PHC clinics were 0.029 and 0.116 kg/visit, respectively (Table 3).

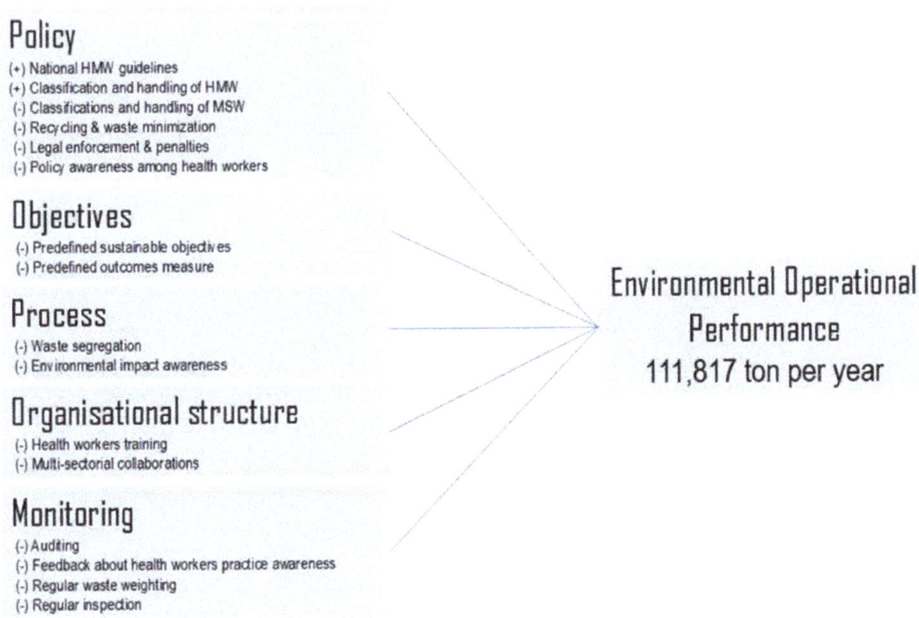

Figure 3. Document analysis results according to Trumpp et al.'s framework. (+): opportunities; (-): challenges.

Table 3. Saudi healthcare facility waste indicators (2018).

Healthcare Facilities	Indicators	
Hospitals	Number of hospitals	284
	Total beds	43,680
	HMW (tons/year)	21,016
	Estimated MSW (tons/year) [a]	84,064
	Estimated HMW (kg/bed/day)	1.711
	Estimated MSW (kg/bed/day)	6.847
PHC clinics [b]	Number of PHC clinics	2198
	Number of visits	53,169,954
	HMW (tons/year)	1347
	Estimated MSW (tons/year) [a]	5390
	Estimated HMW (kg/visit)	0.029
	Estimated MSW (kg/visit)	0.116

[a] Estimation based on WHO categorization. [b] Data from 192 clinics were not available.

To estimate the MSW composition in Saudi Arabia, the waste was calculated in accordance with MSW values from hospitals in Kuwait [33]. In addition to the data source reliability, this study was selected for two reasons. First, Saudi Arabia and Kuwait are similar countries in terms of their health system operational capacity. The ratio of physicians per 1000 population was 2.64 for Saudi Arabia and 2.61 for Kuwait, and that of hospital beds per 1000 population was 2.04 for Saudi Arabia and 2.24 for Kuwait [34]. Second, Kuwait belongs to the Gulf States and is governed by the Healthcare Waste Act, in line with other GCC countries, so this study was considered appropriate for guiding our estimations [25] (Table 4).

Table 4. Estimated amount of MSW in Saudi Arabia hospitals in accordance with Hamoda and Bahman's study [33].

Components	Percentage Content (%) in Kuwait (Hamoda and Bahman, 2007) [33]	Estimated Waste Mass in KSA (tons/year)
Paper and cardboard	32	26,900
Plastic	18	15,131
Glass	10	8406
Metal	9	7565
Food	12	10,087
Textiles	11	9247
Others	8	6725
Total	100	84,064

3.2. *Meso Level: Healthcare Organization*

Observations and interviews of key informants revealed sufficient budgetary allocation for HMW management, where waste was collected and segregated in different types of receptacles by dedicated staff. However, the hospital lacked a clear policy. Furthermore, as confirmed previously in document analysis, training only covered hazardous waste handling, and there was no form of color coding and segregation of the MSW. Due to the unclear policy and inadequate sustainable waste handling guidelines, healthcare facilities lacked monitoring, even as our observation confirmed that waste weighing was not regularly conducted.

With regard to recycling, direct observations, interviews, and the focus group showed that there was no form of MSW recycling or reuse. All wastes were treated by a purpose-built waste treatment plant and disposed of in engineered landfill sites, where the waste was handled by alternative technology. According to the participants, waste recycling was identified as a primary opportunity for the Saudi national system. One of the participants at a high senior position in the MOH said:

"The Kingdom is heading towards achieving sustainable development. Recycling is the future trade for Saudi Arabia."

Several other participants shared the same view about healthcare waste recycling:

"Waste is a fortune, and proven by global experiences, as it is a great economic return. If we recycle waste, we will achieve two goals: reduce burial and recover part of the cost of these materials."

"Large industries will be based on recycling, and will reduce the cost of waste collection, transportation and disposal."

According to our rapid assessment scores achieved by triangulating our observations with qualitative answers to ensure consistency, the level of sustainability of the hospital waste management system was considered at level 2. This means that the healthcare waste management system of the hospital was operating in a manner consistent with some aspects considered sustainable in terms of waste storage and disposal and other aspects considered unsustainable, especially in waste recycling, as seen in Table 5.

Table 5. Healthcare waste management practices in the case study hospital.

Waste Management Indicators	Sustainability Level
General management strategy	
Hospital waste management policy or strategy	3
Special budget for waste management	4
Operative staff for waste management	4
Training on waste management	2
Waste collection and segregation	
Number of containers	4
Type of containers	4
Color coding	2
Waste collection and segregation	2
Quantity of waste	2
Waste storage	
Washing and disinfecting of storage containers	3
Separate packaging and transport of waste	4
Waste stored in areas with built-in roofs	3
Waste recycling	
What is recycled?	0
What is reused?	0
Waste treatment (off-site)	
Waste treatment in a purpose-built waste treatment plant	4
Autoclaving of lab waste	3
Chemical disinfection of body fluids	3
Incineration outside	3
Encapsulation of sharps	3
Off-site disposal	
Transport in enclosed vehicles	4
Final destination	3

3.3. Micro Level (Individuals)

The interviews and focus group revealed that the participants identified social responsibility as a main challenge in achieving environmental sustainability. Table 6 presents the identified dimensions of social responsibility.

Table 6. Overview of identified social responsibility dimensions.

Theme	Subthemes
Social responsibility	Awareness Education Society/culture Information dissemination Multi-sectorial collaboration

The participants believed that the Saudi policymakers faced a huge challenge in achieving sustainable waste management, and they suggested that there was a need to strengthen social responsibility through education and increased awareness among healthcare workers and the community. One of the participants in an academic job said:

"Through my academic research, I found that the level of social responsibility here in Saudi Arabia in dealing with waste in general is unsatisfactory."

The participants believed that the need for awareness was a key factor for enhancing social responsibility, as said by one of the medical waste workers:

"We will achieve the optimal level of social responsibility if the society is aware of the great economic return and the clean environment that comes from sustainable waste practice."

The Saudi policymakers recognized the need to increase the value of social responsibility in the community. Document analysis showed that the Saudi Ministry of Environment recently updated the general environmental regulation in 2020 and added the following statement: "Raise the awareness of environmental issues, instill a sense of individual and collective social responsibility to preserve and improve it, and encourage national volunteer effort in this field" [26].

Raising awareness was strongly emphasized in the documents analyzed, and one of the participants suggested:

"The most important challenge is the culture of society, the information dissemination methods must be changed, dependence on educational posters is insufficient, and so it must attract the individual's eye through the channels that he follows such as social media platforms."

Suggestions for healthcare waste management practices were not limited to healthcare providers; one of the environmental department managers complained:

"Medical waste does not exist only inside of healthcare institutions. Do not forget that there is dangerous hazardous waste in homes and this must be disposed of in a safe way. Unfortunately, we do not have strict legislation to sort the household waste and dispose of medical waste in a safe way."

Generally, the main challenge toward sustainable healthcare waste practices was individual and community awareness and a sense of social responsibility. Document analysis and interviews showed that the government still faces key challenges in achieving its new environmental goals. Raising social awareness requires the cooperation of various sectors as well as the sharing of information between the parties concerned. The focus group concluded that the dissemination of information about the environmental and financial benefits of sustainable practices could motivate individuals toward more positive sustainable behaviors.

4. Discussion

To the best of our knowledge, this was the first study to provide a comprehensive picture of the sustainability of healthcare waste management in Saudi Arabia. Adopting a multi-method approach combining primary and secondary data showed a lack of studies on the subject from a more holistic perspective. While nine national and regional policy documents mentioned healthcare waste, this study found that these documents were limited in scope, as only one document addressed four of five environmental sustainability framework elements [29]. Overall, the documents largely ignored the fact that the sustainability of healthcare waste management has specific features regardless of whether it is related to policies, organizations, or individuals [23,25,27,28,30–32,35]. There are a number of challenges facing the sustainability of the Saudi waste management system, including the lack of a legal framework, limited training on sustainable management practices, a lack of coordination among key stakeholders, and the absence of social responsibility. All these challenges pose significant obstacles to the waste management system's sustainability in Saudi Arabia and are linked to the macro, meso, and micro levels.

At the national level, Saudi Arabia's healthcare facilities produce a large quantity of healthcare waste and consume just as much natural materials. The country's estimated HMW and MSW were 22,000 and 98,000 tons/year, respectively. However, these figures seemed acceptable for Saudi Arabia as they were the same as those for other countries with similar economic conditions [6,36]. With regard to the consumption of natural resources, Saudi government hospitals across the country, every year, throw away in landfills paper (27,000 tons), plastic (15,000 tons), food (10,000 tons), glass (8000 tons), and metal (7000 tons). Regrettably, all these tons of materials end up in landfills without any form of recycling. From the perspective of sustainability, it seems that Saudi Arabia, through its advanced landfill HMW management system, is moving toward achieving enhanced life on land (SDG15) and good health and well-being (SDG03), two of the Sustainable Development Goals. However, there are still opportunities for stakeholders to meet the challenges in managing MSW and reaching responsible consumption and production (SDG12). By so doing, the country will achieve economic growth (SDG08), which is another main Sustainable Development Goal [9]. From a political point of view, it is important to develop a coordinated policy that balances the economic, social, and environmental dimensions of the healthcare waste management system [37]. This can be achieved by creating a legislative organization that mobilizes and unifies the practices of all the parties involved in the healthcare waste industry in order to create a shared vision for sustainable development in the country.

Much of the national unsustainability of the healthcare waste system can be traced to matters relating to policy, objectives, processes, organizational structure, and environmental monitoring [19]. Internationally, healthcare waste is regulated either through ordinance or a specific policy [38]. According to Ali et al. and other researchers, a specific policy for healthcare waste is considered a recent phenomenon in most developing countries; therefore, there are still many shortcomings regarding waste practices [6,12,39]. This seems to be the case in Saudi Arabia, too, so a proper waste management policy is needed for the country to achieve economic and environmental sustainability. Our results also emphasize the need for the country to extend its policy beyond healthcare institutions. In addition, the national policy objectives must be developed within the context of sustainability and in a comprehensive manner that will align the mission and vision statements of healthcare organizations with those values entrenched at the organizational level. With regard to the process, we noted an inconsistency between the highly clarified HMW policy at the macro level and operational process at the meso level. Although the policy demands that HMW be collected separately and be regulated under GCC hazardous waste regulations (Articles 4 and 5), both document analysis and observations revealed an ignorance of the policy and procedures [27–32]. However, healthcare waste malpractice is not limited only to Saudi Arabia but is also noted in other developed and developing countries [6,11,12,39,40].

All the above-mentioned inconsistencies might be due to the lack of an organizational structure in terms of a merger of the political will, public engagement, and multi-sectorial collaboration [23]. Saudi policymakers recognize that the gap in translating policy into practice is the result of the hierarchical waste governance system in the country. Therefore, the government recently established the SWMC as a new institutional framework to organize and make legislation for the management of all types of wastes across all parties involved, including different ministries and organizations. The presence of this new legislative institution opens the door for innovative initiatives that could lead the medical waste management process toward a more sustainable performance. This would then ensure the cooperation of all stakeholders, the health of the environment, and the strengthening of the role of community. In fact, the social dimension deserves more attention, as seen in a developing country like Brazil, where Scavard and colleagues proposed a new healthcare waste management framework tailored to the country's cultural and economic context. This required establishing educational programs and empowering the community in its role. This initiative not only resulted in the enhancement of environmental management sustainability but also improved the population's quality of life [41].

With regard to monitoring of environmental issues, although the MOH invested in the technical and financial aspects of the process as well as staffing and landfilling procedures, waste training and recycling received scant attention among healthcare workers. With the attention focused mainly on waste personnel, it is crucial that all healthcare providers be involved. For example, in Vietnam, the involvement and commitment of all healthcare workers through the training programs achieved positive results in terms of the overall monitoring system [42]. It seems that that the Saudi healthcare waste management system is still based on the traditional linear economic model of take–make–consume–dispose. This is considered an obstacle toward achieving sustainability, due to its negative economic, environmental, and social impact [43]. Transition toward a circular economy by implementing closed loops (reduce, reuse, and recycle) can positively affect the status of the environmental management system, as reported previously [8].

To enhance the environmental sustainability level of healthcare organizations, the implementation of a healthcare waste policy should involve all levels of staff, including healthcare workers. Furthermore, since the healthcare waste pattern varies among departments, training should be designed according to specific staff and departmental requirements [44,45]. The provision of training programs in environmental management is a key factor that should encourage the staff toward a more environmentally friendly attitude and practice. However, relying only on a passive training approach does not always bring about long-term behavioral changes [46]. Enhancing enablers and eliminating barriers to proper waste management can also help create positive behavioral changes. Alzahrani suggested that the healthcare waste management policy in Saudi Arabia must be based on both a bottom-up and a top-down approach, where staff training and staff empowerment to tackle waste represent the bottom-up solution, while law and regulations aligned with sustainable waste practices represent the top-down solution [23].

At the individual level, some of the participants did not view sustainability as their social responsibility or waste management as part of their jobs; rather, they suggested that the main factors that affected employee waste management practices were linked to the organizational culture. Therefore, any attempts to introduce recycling and other sustainable practices need to be incorporated into the healthcare waste policy. Transparency, in terms of the publication of sustainability reports, was also identified as a crucial factor affecting social responsibility [47]. Sharing of information about the economic and environmental benefits of sustainable waste practices would make the employees more likely to become environmentally friendly. Therefore, health workers should be made aware of the financial cost of and carbon emissions contributed by each type of waste. The involvement of healthcare workers in developing and implementing a waste policy would also increase their sense of ownership, which, in turn, would make them more likely to be responsible for the successful implementation of any sustainable measures [4,44].

5. Conclusions

This study explored the sustainability of the current Saudi healthcare waste management system from different perspectives. The major critical points were identified based on an examination of the information obtained from document analysis, interviews of key informants, a focus group session, and observations. Despite the heavy financial investment in HMW management, the MSW management system is not well equipped. Consequently, there is an ever-increasing load on the system, resulting in an unsustainable healthcare waste management system. Generally speaking, sustainable healthcare waste management practices can be achieved by integrating an element of social responsibility within the waste policy as well as by enhancing employee awareness, adding a training component, developing programs, and increasing the promotion of the environmental and economic benefits of sustainable practices.

The process of working through data triangulation was considered effective as the combination of primary and secondary sources not only strengthened data reliability and validity but also provided a holistic view to better inform further investigation. However, the study had some limitations, as our observations were based on subjective assessment and, therefore, more evidence was needed. In addition, HMW and MSW estimation was not especially reliable due to the shortage and poor quality of data in some health facilities. However, we believe that our estimations give a rough indication of the total healthcare waste produced and the natural resources consumed. To the best of our knowledge, our extensive document analysis revealed that there is only one study in all the GCC countries that assesses MSW in healthcare facilities [33]. Therefore, future research is required to estimate HMW and MSW production and components. Overall, despite the limitations, this paper should facilitate the development of a waste policy and some of the actions that healthcare organizations can take toward the creation of a more environmentally sustainable waste management system. These actions might include future plans that enhance documentation, digitization, and effective multisectoral collaboration. The results can be generalized to other developing regions, especially the GCC countries, which are similar not only in their healthcare waste policies and procedures but also in terms of their economic growth, culture, and community.

Author Contributions: N.S.A.; Conceptualization; Project administration; Methodology, reviewing and editing the final draft. M.Y.Q.; Data collection; Reviewing and editing, J.H.A.; Data collection; Validation; Reviewing and editing. All authors have read and agreed to the published version of the manuscript.

Funding: This research received no external funding.

Institutional Review Board Statement: The study was conducted according to the guidelines of the Declaration of Ethics Committee at King Saud University (protocol code KSU-HE-19-287 and date of approval 28/7/2019).

Informed Consent Statement: Informed consent was obtained from all subjects involved in the study.

Data Availability Statement: Not applicable.

Acknowledgments: The authors extend their appreciation to the Research Center for Humanity, Deanship of Scientific Research at King Saud University, for funding this research Group No (HRGP-1-19-05).

Conflicts of Interest: The authors declare no conflict of interest.

Appendix A

Table A1. Quantity of HMW and MSW according to Saudi Arabia's regions.

Region	Number of Hospitals	Total Beds	HMW (kg/year)	Estimated MSW (kg/year)
H Riyadh	49	8337	2,465,309	9,861,236
Makkah	10	2694	1,725,223	6,900,892
Madinah	19	2786	1,398,441	5,593,764
Qassim	19	2859	712,781	2,851,124
Eastern Province	20	3366	4,829,084	19,316,336
Jeddah	13	3091	2,213,884	8,855,536
Jizan	21	2225	1,610,649	6,442,596
Najran	10	1300	531,732	2,126,928
Asir	20	2330	1,151,984	4,607,936
Bahah	10	1165	525,899	2,103,596
Taif	16	2640	450,934	1,803,736
Ha'il	13	1790	194,528	778,112
Jawf	9	1330	534,514	2,138,056
Ahsa	10	1955	404,901	1,619,604
Northern Borders	10	1360	369,356	1,477,424
Hafer Albaten	7	1000	164,087	656,348
Tabuk	12	1820	942,660	3,770,640
Bishah	7	770	213,845	855,380
Qurayyat	4	490	334,122	1,336,488
Qunfudah	5	400	242,089	968,356
Total	284	43,680	21,016,022	84,064,088

Table A1. *Cont.*

Region	Number of Hospitals	Total Beds	HMW (kg/year)	Estimated MSW (kg/year)
Riyadh	447	7,624,088	473,648	1,894,592
Makkah	85	3,262,169	NA	NA
Madinah	159	5,448,687	55,598	222,392
Qassim	183	3,517,693	54,764	219,056
Eastern Province	143	3,196,355	NA	NA
Jeddah	95	3,262,169	33,908	135,632
Jizan	170	4,507,225	145,641	582,564
Najran	68	1,505,813	3558	14,232
Asir	254	3,727,489	137,814	551,256
Bahah	108	1,617,888	33,980	135,920
Taif	121	2,047,850	27,130	108,520
Ha'il	111	1,897,250	64,969	259,876
Jawf	43	1,035,068	185,105	740,420
Ahsa	72	3,300,629	NA	NA
Northern Borders	48	1,060,549	96,911	387,644
Hafer Albaten	40	954,542	4367	17,468
Tabuk	96	1,502,303	19,804	79,216
Bishah	85	1,216,468	10,324	41,296
Qurayyat	19	294,253	NA	NA
Qunfudah	43	997,493	NA	NA
Total	2390	46,220,553	1,347,521	5,390,084

References

1. Jameton, A.; McGuire, C. Toward sustainable health-care services: Principles, challenges, and a process. *Int. J. Sustain. High. Educ.* **2002**, *3*, 113–127. [CrossRef]
2. Seifert, C.; Guenther, E. Prevention is better than cure-Environmental management measures in hospitals. *Corp. Soc. Responsib. Environ. Manag.* **2019**, *26*, 781–790. [CrossRef]
3. Dace, E.; Stibe, A.; Timma, L. A holistic approach to manage environmental quality by using the Kano model and social cognitive theory. *Corp. Soc. Responsib. Environ. Manag.* **2020**, *27*, 430–443. [CrossRef]
4. Duane, B.; Ramasubbu, D.; Harford, S.; Steinbach, I.; Swan, J.; Croasdale, K.; Stancliffe, R. Environmental sustainability and waste within the dental practice. *Br. Dent. J.* **2019**, *226*, 611–618. [CrossRef]
5. World Health Organization. Definition and Characterization of Health-Care Waste. 2017. Available online: https://www.who.int/water_sanitation_health/medicalwaste/002to019.pdf (accessed on 27 November 2020).
6. Ali, M.; Wang, W.; Chaudhry, N.; Geng, Y. Hospital waste management in developing countries: A mini review. *Waste Manag. Res.* **2017**, *35*, 581–592. [CrossRef]
7. Zimmer, C.; McKinley, D. New approaches to pollution prevention in the healthcare industry. *J. Clean. Prod.* **2008**, *16*, 734–742. [CrossRef]
8. Fonseca, L.M.; Domingues, J.P.; Dima, A.M. Mapping the Sustainable Development Goals Relationships. *Sustainability* **2020**, *12*, 3359. [CrossRef]
9. United Nations (UN). Transforming Our World: The 2030 Agenda for Sustainable Development. 2015. Available online: https://sustainabledevelopment.un.org/content/documents/21252030%20Agenda%20for%20Sustainable% (accessed on 27 January 2021).
10. Alharbi, N.S.; Qattan, M.Y.; Alhaji, J.H. Towards Sustainable Food Services in Hospitals: Expanding the Concept of 'Plate Waste' to 'Tray Waste'. *Sustainability* **2020**, *12*, 6872. [CrossRef]
11. Caniato, M.; Tudor, T.; Vaccari, M. Understanding the perceptions, roles and interactions of stakeholder networks managing health-care waste: A case study of the Gaza Strip. *Waste Manag.* **2015**, *35*, 255–264. [CrossRef]
12. Caniato, M.; Tudor, T.L.; Vaccari, M. Assessment of health-care waste management in a humanitarian crisis: A case study of the Gaza Strip. *Waste Manag.* **2016**, *58*, 386–396. [CrossRef] [PubMed]
13. United Nations (UN). High-Level Political Forum. 2018. Available online: https://sustainabledevelopment.un.org/hlpf/2018 (accessed on 15 March 2020).
14. World Atlas. Saudi Arabia. 2015. Available online: https://www.worldatlas.com (accessed on 15 March 2020).
15. MOH. Ministry of Health. Statistical Yearbook. 2018. Available online: https://www.moh.gov.sa/en/Ministry/Statistics/book/Documents/book-Statistics.pdf (accessed on 27 November 2020).
16. General Authority for Statistics. Sixteenth Services Directory 2017 Riyadh Region. 2017. Available online: https://www.stats.gov.sa/sites/default/files/ar-riyad_region_ar.pdf (accessed on 27 November 2020).

17. Cucchiella, F.; D'Adamo, I.; Gastaldi, M. Sustainable waste management: Waste to energy plant as an alternative to landfill. *Energy Convers. Manag.* **2017**, *131*, 18–31. [CrossRef]
18. Barnett-Page, E.; Thomas, J. Methods for the synthesis of qualitative research: A critical review. *BMC Med Res. Methodol.* **2009**, *9*, 59. [CrossRef]
19. Trumpp, C.; Endrikat, J.; Zopf, C.; Guenther, E. Definition, Conceptualization, and Measurement of Corporate Environmental Performance: A Critical Examination of a Multidimensional Construct. *J. Bus. Ethic* **2015**, *126*, 185–204. [CrossRef]
20. Townend, W.K.; Cheeseman, C.R. Guidelines for the evaluation and assessment of the sustainable use of resources and of wastes management at healthcare facilities. *Waste Manag. Res.* **2005**, *23*, 398–408. [CrossRef] [PubMed]
21. Creswell, J.W. *Research design: Qualitative, Quantitative, and Mixed Method Approaches*, 4th ed.; International Student Edition; Sage Publications: Los Angeles, CA, USA, 2014.
22. Brewer, J.; Hunter, A. *Multimethod Research: A Synthesis of Styles*; Sage Publications, Inc.: Los Angeles, CA, USA, 1989.
23. Alzahrani, D. Environmental Legislation and the Management of Medical Waste. In *People and the Planet 2013 Conference Proceedings*; Global Cities Research Institute, RMIT University: Melbourne, Australia, 2013; pp. 1–8.
24. Middle East Business Intelligence. Saudi Arabia: Government Guidelines May Boost Demand for Medical Waste Treatment Equipments. 2002. Available online: https://www.meed.com/countries/saudi-arabia?location=4913 (accessed on 8 November 2020).
25. The Gulf Cooperation Council. The Executive Regulations of Healthcare Waste Management That Endorsed by the Gulf Cooperation Council (Issue 275). 2019. Available online: www.moh.go.sa (accessed on 15 January 2020).
26. Saudi Press Agency (SPA). The Saudi Waste Management Center. 2020. Available online: https://www.spa.gov.sa/viewstory.php?lang=ar&newsid=2109969 (accessed on 27 November 2020).
27. Al-hadlaq, A.; Ali, Z.; Balachandran, W. Bio-Medical Waste Handling and Management in Riyadh, Saudi Arabia. *Int. J. Chem. Environ. Eng* **2013**, *4*, 343–352.
28. Aljabre, S.H.; Hoffmann, F.; Almorzog, B.S.; Mikiling, L.; Alabdulatif, M.; Al-Quorain, A.A. Hospital gener-ated waste: An assessment of the awareness of hospital staff. *J. Fam. Community Med.* **2002**, *9*, 47–50.
29. Bdour, A.N.; Tarawneh, Z.; Al-Momani, T.; El-Mashaleh, M. Analysis of hospital staff exposure risks and awareness about poor medical waste management-A case study of the Tabuk regional healthcare system-Saudi Arabia. *J. Commun. Dis.* **2015**, *47*, 1–13.
30. Habeeb, T.H.; Ahmad, S. Handling Health Care Waste Management and gender differences in the Madinah Primary Healthcare Centers, Kingdom of Saudi Arabia. *Malays. J. Soc. Space* **2015**, *6*, 47–55.
31. Omer, E.; Alsubaie, A. Medical Waste Management Survey in Primary Health Care Centers, Saudi Arabia. *Majmaah J. Health Sci.* **2017**, *7*, 7–22. [CrossRef]
32. Aljabre, S.H. Hospital generated waste: A plan for its proper management. *J. Fam. Community Med.* **2002**, *9*, 61–65.
33. Hamoda, H.M.; El-Tomi, H.N.; Bahman, Q.Y. Variations in Hospital Waste Quantities and Generation Rates. *J. Environ. Sci. Health Part A* **2005**, *40*, 467–476. [CrossRef] [PubMed]
34. World Bank. Health Indicators. 2019. Available online: https://data.worldbank.org/indicator/SH.MED.NUMW.P3 (accessed on 27 November 2020).
35. Alagha, O.; AlOmari, A.; Jarrah, N. Medical waste management and production rate in the Eastern Province of the Kingdom of Saudi Arabia. *Euro-Mediterr. J. Environ. Integr.* **2018**, *3*, 35. [CrossRef]
36. World Health Organization. *Waste from Health-Care Activities*; Fact Sheet no. 253; World Health Organization: Geneva, Switzerland, 2011; Available online: http://www.who.int/mediacentre/factsheets/fs253/en (accessed on 27 November 2020).
37. Fonseca, L.M.; Domingues, J.P.; Pereira, M.T.; Martins, F.F.; Zimon, D. Assessment of Circular Economy within Portuguese Organizations. *Sustainability* **2018**, *10*, 2521. [CrossRef]
38. .Mühlich, M. Comparison of infectious waste management in European hospitals. *J. Hosp. Infect.* **2003**, *55*, 260–268. [CrossRef]
39. Ciplak, N.; Kaskun, S. Healthcare waste management practice in the West Black Sea Region, Turkey: A comparative analysis with the developed and developing countries. *J. Air Waste Manag. Assoc.* **2015**, *65*, 1387–1394. [CrossRef]
40. Hangulu, L.; Akintola, O. Health care waste management in community-based care: Experiences of community health workers in low resource communities in South Africa. *BMC Public Health* **2017**, *17*, 448. [CrossRef]
41. Scavarda, A.; Daú, G.L.; Scavarda, L.F.; Korzenowski, A.L. A proposed healthcare supply chain management framework in the emerging economies with the sustainable lenses: The theory, the practice, and the policy. *Resour. Conserv. Recycl.* **2019**, *141*, 418–430. [CrossRef]
42. Kühling, J.-G.; Pieper, U. Management of healthcare waste: Developments in Southeast Asia in the twenty-first century. *Waste Manag. Res.* **2012**, *30*, 100–104. [CrossRef]
43. Segerson, K.; Pearce, D.W.; Turner, R.K. Economics of Natural Resources and the Environment. *Land Econ.* **1991**, *67*, 272. [CrossRef]
44. Tudor, T.; Noonan, C.; Jenkin, L. Healthcare waste management: A case study from the National Health Service in Cornwall, United Kingdom. *Waste Manag.* **2005**, *25*, 606–615. [CrossRef] [PubMed]
45. Gao, Q.; Shi, Y.; Mo, D.; Nie, J.; Yang, M.; Rozelle, S.; Sylvia, S. Medical waste management in three areas of rural China. *PLoS ONE* **2018**, *13*, e0200889. [CrossRef] [PubMed]
46. McKenzie-Mohr, D. *Fostering Sustainable Behavior: An Introduction to Community-Based Social Marketing*; New Society Publishers: Gabriola Island, BC, Canada, 2011.
47. Falcone, P.M.; García, S.G.; Imbert, E.; Lijó, L.; Moreira, M.T.; Tani, A.; Tartiu, V.E.; Morone, P. Transitioning towards the bio-economy: Assessing the social dimension through a stakeholder lens. *Corp. Soc. Responsib. Environ. Manag.* **2019**, *26*, 1135–1153. [CrossRef]

sustainability

MDPI

Article

An Overview on Solid Waste Generation and Management: Current Status in Chile

Romina Cayumil [1], Rita Khanna [2,*], Yuri Konyukhov [3], Igor Burmistrov [4], Jumat Beisembekovich Kargin [5] and Partha Sarathy Mukherjee [6]

1 Facultad de Ingenieria & Centro de Investigación Para La Sustentabilidad, Universidad Andres Bello, Santiago 7500000, Chile; rcayumil@gmail.com
2 School of Materials Science and Engineering (Ret.), The University of New South Wales, Sydney, NSW 2052, Australia
3 Department of Functional Nanosystems and High-Temperature Materials, National University of Science and Technology "MISiS", Moscow 119049, Russia; martinsit@mail.ru
4 Engineering Centre, Plekhanov Russian University of Economics, Moscow 117997, Russia; glas100@yandex.ru
5 Department of Technologies Commercialization, L.N. Gumilyov Eurasian National University, Nur-Sultan 010008, Kazakhstan; kjb_orken@mail.ru
6 Institute of Minerals and Materials Technology (Ret.), Council of Scientific and Industrial Research, Bhubaneshwar 751013, India; psmukherjee52@gmail.com
* Correspondence: rita.khanna66@gmail.com; Tel.: +61-0434155956

Abstract: The widespread generation of, ever increasing volumes of and the sustainable management of solid wastes are global issues of great concern. Due to wide variations in composition and associated complexities, significant efforts are required for their collection, processing and environmentally safe disposal in a cost effective manner. An overview of solid wastes is presented in this article with a specific focus on municipal solid wastes and industrial waste from the iron/steelmaking and aluminium industries. Key waste issues such as its sources, compositions, volumes, the factors affecting waste generation and waste processing are first discussed, followed by a further discussion regarding recycling, resource recovery, disposal and the associated environmental impacts. In a special case study, waste generation and management in Chile is presented in greater detail. Detailed information is provided on government initiatives and legislation for integrated solid waste management and its movement towards a circular economy. Measures include regulations on waste management framework which concerns the transboundary movements of hazardous wastes, persistent organic pollutants, the closure of mining activities and installations and restrictions on plastics disposal. With Chile being world's largest producer of copper, significant efforts for mining waste management, its infrastructure and procedures are being put in place to reduce the environmental impact of the mining sector and its associated waste generation.

Keywords: solid wastes; MSW; industrial wastes; iron/steel sector; aluminium sector; copper mining

Citation: Cayumil, R.; Khanna, R.; Konyukhov, Y.; Burmistrov, I.; Kargin, J.B.; Mukherjee, P.S. An Overview on Solid Waste Generation and Management: Current Status in Chile. *Sustainability* **2021**, *13*, 11644. https://doi.org/10.3390/su132111644

Academic Editor: Vincenzo Torretta

Received: 14 September 2021
Accepted: 12 October 2021
Published: 21 October 2021

1. Introduction

Solid waste management is a global issue affecting individuals, communities and governments in most countries. Poor and inadequate waste management decisions can affect daily health, cleanliness and productivity, thereby affecting economic development at all levels of society. Some of the issues include the contamination of oceans, flooding, clogged drains, respiratory diseases, airborne particulates, waste consumption by animals and air and land pollution. The growing level of prosperity, rapid urbanization and population growth has led to increases in the per capita generation of waste. In addition, waste management is often administered by local authorities with limited capacity for management, operational monitoring and financial resources. It has been estimated that every year, over a billion tonnes of waste are produced globally. This number continues to rise for most nations with data reported in 2018 at: East Asia and the Pacific (468 MT); Europe and Central Asia (392 MT); South Asia (334 MT); North America (289 MT); Latin America (231 MT); Sub-Saharan Africa (174 MT) and Middle east and North Africa (129 MT) [1,2]. Unless urgent action is taken on several waste management fronts, global waste generation rates may increase to 20 billion tonnes/year by 2050.

Due to the wide variation in waste composition and characteristics, different types of solid wastes can be classified as: Municipal Solid Waste (MSW), Industrial Waste (IW), Agricultural Waste (AW), Construction and Demolition Waste (C&DW), Hazardous Waste, Medical Waste and Electronic Waste (e-waste). The world-bank has reported that high-income countries, with a population of 16% of globe inhabitants, produce approximately 34% of the planet's waste [3]. Table 1 summarizes the amount of different wastes generated per capita per day worldwide. The largest quantities of wastes were found to be industrial types of waste (12.73 kg), followed by agricultural waste (3.35 kg). Further, there was an estimated 1.68 kg of C&D waste, 0.74 kg of MSW produced globally, followed by 0.32 kg of hazardous waste, 0.25 kg of medical waste, and 0.02 kg of e-waste [4].

Table 1. Global generation of various wastes.

Type of Waste	Generation per Capita (kg/Capita/Day)
Industrial waste	12.73
Agricultural waste	3.35
Construction and demolition waste	1.68
Municipal solid waste	0.74
Hazardous waste	0.32
Medical waste	0.25
Electronic waste	0.02

Through a focus on solid waste generation and management, this article is organized as follows: an overview of different types of wastes is presented in Section 2 along with detailed statistics on two major solid wastes, namely, municipal solid waste and industrial waste from key metal processing industries. The complexities associated with these wastes including their composition, generation, collection, volumes, recycling methods, disposal and environmental impact will be discussed in the global scenario. Section 3 focusses on the practical realities of solid waste management in Chile, serving as a representative example. With a population of over 18 million and a total area 756,000 km^2, this South American country borders the South Pacific Ocean to the west with a coastline of over 6000 km long. Around 90% of the population lives in urban areas, primarily in the metropolitan area of Greater Santiago. Chile is among Latin America's fastest-growing economies in recent decades, enabling the country to significantly reduce poverty. Chile is gradually moving to a circular economy through legislation, long-term plans and the provision of funds from public and private entities and has demonstrated a financial implementation of these objectives.

2. Generation, Management and Disposal of Solid Wastes: A global Overview

2.1. *Municipal Solid Waste (MSW)*

2.1.1. Sources and Composition

Municipal solid waste represents the typical waste generated by households, offices, commercial shops, hotels, schools and other institutions. The major contributors are food waste, paper, plastics, metal, yard waste, cardboard and packing waste glass. It can also include some demolition and construction debris along with small amounts of hazardous and chemical waste, e.g., electric light bulbs, batteries, automotive parts, discarded medicines and chemicals. An overview of the typical sources of MSW and its constituents are given in Table 2 [5].

Table 2. Typical sources and compositions of MSW.

Waste Sources	Waste Generators	Typical Wastes
Residential	Single and/or multi-family dwellings, apartment blocks, high-rise buildings.	Food wastes, paper, cardboard, plastics, yard wastes, furniture, glass, metals, bulky items, consumer electronics, whitegoods, batteries and household hazardous wastes
Commercial institutes, shopping malls and organizations	Shops, stores, hotels, restaurants, markets, offices, schools, hospitals, government organisations	Paper, cardboard, plastics, food wastes, glass, wood, special wastes, metals, hazardous wastes
Municipal services	Street cleaning, landscaping, parks, other recreational areas, beaches, wastewater treatment	General wastes from parks, street sweepings, landscape and tree trimmings, other recreational areas, beaches, sludge.

Typical proportions (wt. %) of different MSW components are estimated as: Food waste (25–35); Paper (25–35); Plastics (7–10); Ferrous metals (3–5); Non-ferrous metals (0.5–2); Glass (5–10); Yard waste (10–15); Hazardous waste (1–2) [6]. In middle- and low-income countries, organic waste accounts for more than 50% of the total MSW generated. In high income countries, this percentage reaches 32%, approximately. Recyclable materials range from 10% in low income countries and up to 50% in high income countries. Recovery and recycling of wastepaper has seen a significant increase among the recyclables over time. In high-income countries, one third of the total amount of waste produced is presently recycled or composted [7]. Table 3 summarizes the generation of MSW for different global regions along with projected increases by 2050.

Table 3. Generation of MSW by region (2016), and their projection by 2050 [3].

Region	Generation of MSW (2016) (Million Tonnes)	Projection (2050) (Million Tonnes)
East and Asia Pacific	468	714
Europe and Central Asia	392	490
South Asia	334	661
North America	289	396
Latin America and the Caribbean	231	369
Sub-Saharan Africa	174	516
Middle East and North Africa	129	255

In 2016, the largest amounts of MSW were generated in the East Asia and Pacific Region, with the region generating 468 million tonnes, representing 23% of the total amount of MSW. Waste generation from Sub-Saharan Africa is expected to increase three times by 2050, reaching 516 million tonnes, followed by South Asia with 661 million tonnes by 2050, doubling on the 334 million tonnes produced in 2016. The Middle East and North Africa region will also double their waste by 2050, with an estimated amount of 255 million tonnes. The problem here not only lies with continuously increasing waste volumes, but also with inadequate waste disposal and management [7].

2.1.2. Factors Influencing MSW Generation

Several factors that affect the volume of MSW that is generated includes population growth, income levels, immigration, industrial and economic development. While population growth is considered to be a critical factor influencing MSW generation [8], EPA has also highlighted income and tax revenue as key factors; electricity consumption was another compelling factor [9,10]. Sukholthaman et al. found that MSW generation tended to increase with increasing levels of gross domestic product (GDP) and the consumer purchase index [11]. Liu et al. have reported on the impact of consumption patterns on the generation of MSW in China [12]. These researchers investigated the impact of economic development and consumption levels on MSW generation. The highest levels of MSW generation were found to exist in medium to low economic consumption groups. Developing regions with large development potential had fast economic growth rates and generated more MSW compared with developed regions with marginal growth in MSW generation. These findings have been summarized in Table 4 [13].

Table 4. Typical factors influencing MSW generation.

Global Factors	Local Factors
Demographics	Population, population density, age, gender, occupation, expenditure on groceries; household count, household density, household type and size, income level, energy consumption.
Economics	Gross domestic product, consumer price index, economic growth, employment/unemployment and waste budget.
Technology	Manufacturing standards, engineering issues, inefficient equipment and facilities.
Legislation & Administration	Policies, laws, enforcement level, management institutions, disposal fees, recycling programmes and amounts recycled
Social	Awareness, literacy levels, public cooperation, cultural practices, urbanisation, tourist attractions, political stability.
Consumer behaviour	Consumption demand and patterns, cooking activity, lifestyles, disposal patterns.

2.1.3. Waste Collection and Disposal

In urban areas, MSW collection and transport involves its storage at the generation source and at pick-up points, its collection by crews/trucks driving around the neighbourhood and finally its transport to a disposal station. The collection of MSW is difficult, complex and costly, accounting for up to 60-80% of the total solid waste budget of a typical community. In residential areas, the most common collection methods are via the kerb, an alley or through backyard carry. In a kerb or alley service, the residents carry plastic bags and/or containers to the kerb or collection point and collect the empty container at a later point. The collection of MSW occurs once or twice per week in most communities. Wang et al. have reported on the rural solid waste management in China, observing that the rural solid waste collection services, facilities and equipment were often of a lower quality [14]. High costs and the inconsistent employment of waste collection workers were found to be unsustainable, and hindered the overall quality of waste collection services [15].

The various materials are separated either by the users at the source or separated from the mixed refuse at a central processing facility. Material separation at the source involves separating the material into different components followed by their transportation. Waste separation at the source is carried out by using separate bins for the biodegradable fraction and the recyclables and hazardous waste materials. The active household response for waste separation has generally been quite poor. The separation of material in MSW usually occurs at a central facility. The organic material is shredded and processed through air classifiers, which separate the desired components for further processing and recycling and for classification as a material or an energy resource. For handpicking, a conveyor moves the solid waste along a group of workers who collect a few of the larger components by hand. A mechanized material recovery method utilizes shearers to break open the bags and trammel screens to separate cans, glass and other inorganic material. Magnetic and electromechanical systems are used for separating ferrous and non-ferrous metals.

An overview of waste disposal methods in the USA is shown in Figure 1 [16]. It is estimated that 75–80% of total MSW generated is collected and only 22–28% is processed and treated. A significant proportion of the collected waste is often dumped indiscriminately, clogging the drains and sewer systems [17]. Additionally, due to imperfect waste infrastructure and recycling systems, the recycling rates of MSW in developing provinces are low. All these factors suggest the generation of large MSW volumes in the near future in developing provinces. In contrast, the developed regions have a stable economic development rate and high recycling efficiency rates of MSW. From the perspective of waste clearance volumes, the developed regions will generate less MSW than developing provinces in the near future [18].

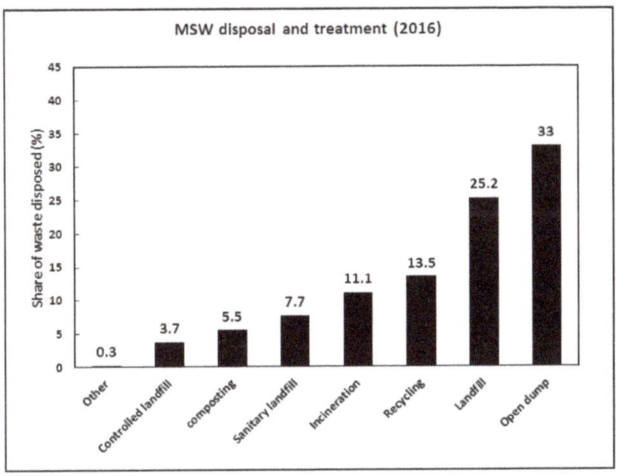

Figure 1. An overview of MSW disposal and treatment in the USA.

2.2. *Industrial Waste*

Industrial activities such as manufacturing processes, mining operations, construction, demolition and the operation of factories produce waste products in the form of materials that have been rendered useless and are no longer required. Industrial wastes include masonry and concrete, dirt and gravel, scrap metal, chemicals, oil, solvents and scrap lumber. Industrial waste can either be solid, semi-solid or liquid in form; they may be hazardous and toxic or may be a non-hazardous waste. Hazardous wastes include, among others explosive wastes, flammable solid wastes, wastes liable to spontaneous combustion, wastes that emit flammable gases on contact with water, oxidising wastes, organic peroxide wastes, infectious wastes and corrosive wastes. [19]. Industrial waste is sometime mixed into municipal waste, making the conduction of accurate assessments difficult. Such waste can pollute the adjacent soil or water bodies thereby contaminating groundwater, rivers or coastal waters [20]. An estimated 7.6 billion tonnes of industrial waste was produced in the USA in 2017. In this overview, we focus our attention on the leading metal processing industries, namely the iron and steelmaking sector and the aluminium industry and the associated waste management.

2.2.1. Iron and Steelmaking Sector

With the ability to adapt to changing conditions regarding raw material availability, energy resources and hot metal demands, ironmaking technology has achieved a high degree of excellence. New challenges, especially with respect to the environment and CO_2 emissions, are the main drivers for technological progress in the 21st century. Steel consumption worldwide is increasing steadily, as is its impact on the environment through energy consumption. Steel provides a key input for other industrial sectors that produce items essential to the functioning of the wider economy, e.g., automobiles, trains, aircrafts, hand tools, complex factory machinery and countless other items.

The industry is continuing to expand, with the application advanced high–strength steels for transport systems, bridges, pylons, construction sector, infrastructure, housing, manufacturing, agriculture and energy [21]. As a permanent material that can be continuously recycled without losing its properties, steel is fundamental to a successful circular economy. The steel industry amounts for 10.7% of national GDP in the USA. Furthermore, China accounts for 31% of steel's value-added contribution, with the USA and Japan accounting for 11% and 10% respectively [22]. The global steel production in 2020 is shown in Table 5, along with the production levels of the top five countries.

Table 5. Global steel production (in million tonnes) in 2020. Contributions from the top five steel producing countries are shown.

Global Production	#1	#2	#3	#4	#5
	China	India	Japan	USA	Russia
1877.5	1064.8	100.3	83.2	72.7	71.6

During the process of ironmaking, the iron ore is converted into different types of iron through several processes (Figure 2). The blast furnace is the most commonly used process and produces pig iron which composed of around 92-94% of iron and 3-5% of carbon along with smaller amounts of other elements. However, pig iron has limited uses, and most of this iron arrives at steel mills, where it is converted into various steel alloys by reducing the carbon content and adding other alloying elements to acquire steel specific properties. Steelmaking uses both recycled materials as well as pig iron from the blast furnace. Both basic oxygen steelmaking (BOS)/ basic oxygen furnace (BOF) and electric arc furnace (EAF) account for virtually all steel production in the world.

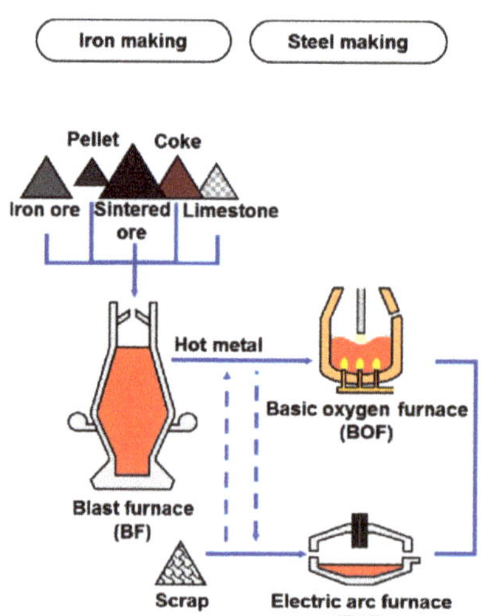

Figure 2. A schematic representation of iron and steelmaking processes.

This industry focuses on reducing raw material consumption and reusing the waste produced in various iron/steelmaking operations. In the 1970/1980s, modern steel plants generally required 1.44 tonnes of raw material to produce 1 tonne of steel; this number has been reduced to ~1.15 tonne of inputs to make 1 tonne of steel as of 2020, thereby achieving a 21%reduction. However, more than a billion tonnes of waste was produced by the iron and steel sector in 2020, largely due to a significant increase in overall steel production. A brief overview of major solid waste products produced during various iron/steel production processes is provided along with techniques used for their disposal in the next section [23–27].

Coke making—These wastes include coal dust, coke dust, tar sludge and acid sludge. Coal dust is recycled through the coal blend and/or by its addition to the sinter charge mix. Tar sludge is added to the coal blend to improve its density. Acid sludge has to be neutralized before its disposal.

Sinter making—The solid wastes generated are dusts and sludges. Dusts are generated at the sinter machine building's floors, in cyclones and the ESP (Electrostatic precipitator). Sludge is generated at the clarifier of the water treatment plant. These materials are recycled back into sinter making, either by blending them with iron ore fines or briquetting.

Pellet plant—The major wastes here are dusts and undersized green pellets. These wastes are generally reused as pellet feed materials.

Blast furnace iron making (BF)—The generated wastes include dusts, burden screenings at the stock house, cast house runner jams, cast house muck, blast furnace slag, flue dust, sludge, refractory waste and hot metal ladle skulls. Dusts and burden screenings are recycled into sinter making by blending them with the iron ore fines in the raw material yard or by being briquetted. Cast house runner jams are either sold as iron scrap for re-melting or recycled in the steel melting shop. Cast house muck is dumped as land-fill material. The air-cooled BF slag is either dumped at landfills or used for road repairs within the steel plant. Refractory waste is reused by the refractory manufacturer in the manufacture of mortars.

Iron making by direct reduction (DRI)—The main solid wastes are char, dusts (both coal and iron) and under sized and reject DRIs. Coal dusts and chars are used in the power plants as fuel for the boiler. Iron dust can also be sold to the cement manufacturers. Under-sized DRIs are briquetted and used in a steel melting shop; DRI rejects are recycled back into the process.

Basic oxygen steel making (BOS)—The main solid wastes include converter slag, scrap, waste refractories, converter muck and gas cleaning plant sludge. Most of these waste materials are produced in a mixed condition, therefore extensive sorting is required prior to their recycling. The plant sludge is dried and blended with the iron ore fines as raw material and are briquetted and reused in sinter making. The converter slag is recycled as a replacement for limestone in the sinter mix. The converter slag can also be used in road repairs, as railway track ballast and as soil conditioner. Scrap and refractory wastes from the converter muck are removed and a balance is used for landfill. There are two types of refractory wastes: magnesium carbon refractories from converter linings, and the other types are from refractory ladles. Alumina based refractory wastes can be used for manufacturing mortars.

Electric Arc Furnace (EAF) steel making—The typical waste products include GCP dust, slag, scrap and waste refractories. Presently, there is not a well-established method for recycling EAF slag. The grinding and processing of EAF is energy intensive due to its high iron oxide content. It has been used for making roads and ceramic tiles.

Other solid waste contributions come from continuous casting, rolling mills and lime calcining, which primarily includes casting scales, refractory scrap, slag, muck, mill scale, limestone and dolomite screenings, lime dust and lime fines.

2.2.2. Aluminium Sector

Currently, aluminium (Al) is the second most widely used metal in the world. To meet the needs and challenges of the 21st century, the production of Al has increased form 1000 tonnes in 1900s to 58 million tonnes in 2016. Being strong, lightweight and durable, Al is increasingly used in the transport sector towards the aim of a partial replacement of steel, which is heavier. A high strength-to-weight ratio means it is particularly useful as a structural material, weighing around 65% less than steel. Lightweight Al also contributes to an increased fuel economy in vehicles from cars to armored tanks. Aluminium is extremely important for the aviation sector, e.g., Airbus A380 uses up to 66%Al in its airframe and a Boeing 747 can use up to 75 tonnes of Al. There is a significant usage of Al in the shipping sector including in military applications, in building and construction, for domes for gymnasiums, schools, storage facilities, industrial roof systems, multi-purpose arenas, packaging, aluminium cans, food containers, wrapping foils, etc. [28]. The global production of aluminium is summarized in Table 6 [29].

Table 6. Global aluminium production (in thousand tonnes) in 2019. Contributions from the top five aluminium producing countries are shown.

Global Production	#1	#2	#3	#4	#5
	China	India	Russia	Canada	U.A.E
64,000	35,800	3700	3600	2900	2700

Bauxite is a key aluminium ore containing ~30–54% alumina (Al_2O_3) along with mixtures of silica, iron oxides, titanium dioxide and several impurities. Around 95% of Al is produced from bauxite ore using the Bayer's process worldwide. In the Bayer process, bauxite is washed in a hot solution of sodium hydroxide, which leaches Al in the form of $Al(OH)_3$; it is later calcined to form Al_2O_3. Red mud (RM), also known as bauxite residue, is a solid waste biproduct of the alumina recovery process. Producing one tonne of alumina consumes up to 2–3 tonnes of bauxite and generates ~0.4–2 tonnes of RM depending on the source/location of the ore; the global average of RM produced is 1.3 tonnes per tonne of alumina [30,31].

Global bauxite resources are estimated to be ~55 to 75 billion tonnes with reserves in Africa (32%), Oceania (23%), South America and the Caribbean (21%) and Asia (18%). With the production of primary aluminium increasing to 47 million tonnes, an estimated 150 million tonnes of RM are being produced in various aluminium plants across the world [32]. The nearly 3 billion tonnes of RM, presently stored in vast waste reserves, represents large industrial waste scenarios around the globe [33]. Several red mud incidents in different countries have also been tabulated in various publications [34].

The Hall–Héroult process is the main process used for smelting aluminium, and involves dissolving alumina in molten cryolite, and electrolyzing the molten salt bath in a purpose-built cell. The electrolysis takes place at 940–980 °C and produces 99.5–99.8% pure aluminium [35,36]. An aluminium smelter produces 40–60 kg of mixed solid wastes per tonne of product, with spent cathodes, spent pot, cell linings comprising the major fraction of waste. The linings consist of 50% refractory material and 50% of carbon. Over the useful life of the linings, the carbon becomes impregnated with aluminium and silicon oxides (averaging 16% of the carbon lining), fluorides (34% of the lining), and cyanide compounds (about 400 ppm). The contamination levels in the refractories element of failed linings are generally low. The other by-products for disposal include skim, dross, fluxing slags, and road sweepings [37–39].

Extensive efforts have been made to process, recycle and reutilize red mud waste. Key challenges in processing RM include high alkalinity (pH: 10–13), small particle sizes (≤75 μm), large volumes and high transportation costs [40,41]. Landfilling and dumping around the industrial plants have been the standard practice worldwide. Vast amounts of stockpiled RM can also be toxic and very hazardous [42]. Deep sea dumping and storage in ponds has also been attempted. Poorly designed storage dams are likely to fail under certain circumstances, resulting in local and environmental contamination. Many efforts are being made to find economically viable and environmentally sustainable solutions to the RM problem. However, most options can only accommodate for a small fraction of the red mud generated globally [43].

The recycling of aluminium requires only 5% of the energy necessary to smelt it from the alumina and a small fraction of greenhouse gases are emitted during aluminium recycling compared with the production of primary aluminium. Properly treated and sorted aluminium scrap can be used to produce different types of goods at low cost given that these scraps already contain some of the alloying elements required for certain applications. Aluminium recycling plants can be found all over the world, but are significant concentrated in Europe, the USA and Japan [44,45].

3. Waste Management in Chile: Current Status

Chile is gradually moving toward a Circular Economy (CE) system. The evolution towards sustainable development and CE is taking place in conjunction with legislation, long-term plans and funds, support and financial implementation from public and private entities. The current national situation regarding waste management and the new *Waste Management Framework* has encouraged the processing of copper mining tailings and prohibited the commercial use of plastic bags. While Chile has a wide variety of resources, mineral resources are the most valuable in economic terms as the national economy is, to a great extent, driven by the mining sector. A wide variety of wastes are generated by the diverse industries in the country. National laws classify waste residues depending on their nature as hazardous waste or non-hazardous waste. The non-hazardous waste is further categorized into industrial waste (IW), municipal solid waste (MSW) and sewage sludge (SS). According to the Chilean Ministry of the Environment, 19.6 million tonnes of total waste were generated in 2018, 97.3% of which were non-hazardous waste. These figures do not include radioactive residues, gangue, low grade minerals, leaching gravel, tailings or slags produced in mineral processing. [46].

This section presents the current status of waste management in Chile and the initiatives towards a more sustainable economy. Waste management activities, waste characterization, regulatory measures, challenges and opportunities and initiatives towards a circular economy will be discussed.

3.1. *Waste Management: Legislations and Regulations*

All countries need to develop a common criteria for the environmental management of waste residues, technical guidelines, regulations with respect to environmental requirements, a holistic management criteria for the minimization of waste generation, and their application during the productive cycle towards clean technologies and circular economy [47]. Chile is a signatory to several international conventions and agreements on waste management and sustainable development. Some of the efforts made by Chile include:

Chile is a member of the OECD, United Nations and The Southern Common Market (MERCOSUR for its Spanish abbreviation), all committed to promote sustainable development. Chile has also subscribed to the Basel Convention as well as to the Stockholm's agreement. Basel Convention establishes mandatory actions for hazardous waste management and appropriate disposal as well as legislation for transboundary movement of waste [48]. The Stockholm convention provides an agreement to protect both human health and the environment from harmful persistent organic pollutants (POP's) [49]. Chile is devoted to eliminating or minimizing the release of POP's to the soil, water and air [50].

The Latin American *"Agreement of Environmental Management of Special Residues and After Use Responsibility"* (Acuerdo de Gestión Ambiental de Residuos Especiales y Responsabilidad Post Consumo), signed during the fourth extraordinary meeting of MERCOSUR's Ministries of the Environment, established the commitment to incorporating production patterns and sustainable consumption, with the objective of minimizing the quantity and harmfulness of waste generated [51]. This agreement also established the term *"special residues of universal generation waste"* based on their environmental impact, hazardous characteristics, risks or potential harmful damage to the environment; such wastes require special environmental management and need to be separated from other residues. Universal wastes include hazardous wastes produced by the households and several businesses including televisions, computers and various electronic devices, batteries, fluorescent lamps, mercury thermostats, mercury containing equipment.

The policy of integrated solid waste management was promulgated in 2005 with the objective to ascertain that all activities related to the management of solid wastes should be carried out with minimal risks for human health and the environment. The following regulatory procedures have been developed for managing and for the disposal of hazardous electronic waste: Law 19,300 provides the *general basis for the environment* [52]; the Basel Convention concerns the transboundary movement of hazardous waste and its elimination; Law 20,920 concerns *Waste Management Framework, Extended Producer Responsibility (EPR)* law [53]; and *promotion for the recycling law, sanitary standard DS 148*, provides regulations for recycling funds [54], among others.

3.2. Waste Management Initiatives

The Chilean government has undertaken several campaigns to increase awareness about recycling. The plan "*Santiago recycles*" began in 2017, and is believed to be one of the most ambitious projects concerning reuse in the country. With a budget of ~5 million euros, the aims of the project were to educate people, build waste management facilities and to make Santiago a more sustainable city [55]. The implementation of the *Waste Management Framework* has provided formal pathways to the producers of lubricants, electric and electronic equipment, batteries, packaging, and rubber tires on their management during the manufacturing process, commercialization, and at their end-of-life, encouraging their use as a resource elsewhere. Extended producer responsibility (EPR) has increased activities to appropriately manage and process these types of wastes. A number of private entities have started businesses for collection and storage, while other companies use these components as B grade raw materials for reuse while minimizing the environmental impact [56].

To avoid the contamination of land and the ocean, a law was approved in 2018 prohibiting the commercial use of plastic bags in the whole territory, with Chile being the first country in Latin America that prohibits plastics use. In 2019, the campaign to reduce single use plastics was launched, particularly plastic straws, and also a campaign to avoid cigarette butts being disposed of in public spaces and/or directly in the soil. The creation of the Circular Economy Office in the Ministry of the Environment was the result of all the actions adopted in waste minimization, management, valorization and recycling [57]. These measures, however, did not include the waste produced by the mining industry. Two key wastes, namely MSW and mining waste, will be discussed in the following sections.

3.3. Municipal and Industrial Wastes

Chile, considered a relatively high-income country, has a population of ~19 million inhabitants. The average generation of MSW per day per capita in Chile is about 1.15 kg, slightly higher than the world's average of 0.74 kg waste per capita [58]. Of the overall amount of MSW produced, 44.9% of MSW was generated in the metropolitan region of Santiago, the capital city of Chile which is home to 41.1% of the country's population. Considerably lower levels of MSW were produced by the Valparaíso, Bío-Bío and Coquimbo regions, accounting for 10.6%, 9.3% and 5.2%, respectively. Approximately 8 million tonnes of MSW is generated in Chile every year; the waste collection rate is almost 100%; the volume of waste rose by 30% during the period 2000–2010.

Household waste is generally collected door-to-door in plastic bags, whereas recyclables are primarily collected at central collection points equipped with containers. Kerb-side collection is still quite limited, with household waste collected on the kerb-side and then transferred into the rear of a compactor truck [59]. The overall efficiency and cost effectiveness is limited as not all citizens are regularly serviced, leading to negative impacts on the environment and the society. The waste is mainly composed of food and green waste, which accounts for 58% of the total waste generated. The remaining waste includes dry recyclables, paper and cardboard (13%), plastics (12%), glass (4%), metal (3%), and up to 15% of organic waste. Approximately 50% of the waste is disposed in sanitary landfills, however, the use of open dumps account for ~27% of disposal and processing.

A further 15% of the MSW is transferred to controlled landfills. The normal collection system transports all the mixed waste to a transfer station or to a landfill [60].

Because they comprise a relatively small fraction of the overall waste, hazardous waste, medical waste and e-waste are generally disposed of along with MSW. Only 5% of the MSW is sent to recycling facilities and less than 1% waste is composted due to limited material separation. Recycling and composting are still emerging treatments and the extent of their usage varies significantly across the country. A new strategy proposes recycling up to 66% of organic waste by 2040, by reducing the level of organic waste produced encouraging its separation in houses, offices, schools and markets, and through the use of infrastructure, equipment and logistic systems that allow the organic residues to be used as a resource in the production of soil enhancers and electric or thermal energy [61,62].

Several methods and recovery systems are currently used in Santiago for recycling the organic fraction of MSW. These include sorted collection at source, separation at source for drop-off collection and sanitary landfills. [63]. Sorting prior to collection at the source involves the sorting of recyclables at home before collection by a municipal service or by informal collectors. Currently, only three municipalities offer a formal sorted collection system; Vitacura and Nuñoa collect dry recyclable waste with a mixed bank collection system, and La Pintana collects kitchen waste with a sorted waste bank. Informal collectors, numbering in excess of 60,000 collect around 2-10 tonnes/month/person of recyclables, and account for ~76% of the total recyclables collected [64]. Project Santiago Recycles introduced the separation of recyclables at the origin for their drop-off collection at solitary bell-shaped container stations associated with charity institutions, and 'clean point' stations with a larger drop-off area for containers of different types of recyclable waste. Recycled waste collected at the largest clean point in Santiago experienced a156% increase between 2009 and 2014. In 2009, there were three sanitary landfills in Santiago and two transfer stations; several dumpsites have been closed in the metropolitan region.

Among the key productive sectors of the Chilean economy are the manufacturing industry, electricity, gas and water supply, livestock industry, agriculture, mining, the wood industry and the fishing industry. As reported by the 'Ministry of Environment' 10.5 million tonnes of non-hazardous industrial waste was generated in 2018 representing ~50% of the total waste produced. The manufacturing industry was the largest waste producer, producing 4.2 million tonnes, followed by the electricity, gas and water supply industry which produced 2.2 million tonnes. The construction industry, real-estate activity sector and commerce respectively produced 0.67, 0.67 and 0.61 million tonnes of waste. The agriculture, livestock, hunting and forestry sectors produced total of 0.61 million tonnes of waste. Smaller amounts of waste were generated by the fishing, small mine & pit, transport, storage & communication sectors, which produced 0.48, 3.61 and 0.3 million tonnes, respectively. A further 0.2 million tonnes of waste were produced by other sectors. Regarding sewage sludge, a 57.8% was generated in the metropolitan Santiago area, which amounted to 198,000 tonnes in 2018. The amount of hazardous waste increased considerably between 2006 and 2018 from 26,840 tonnes (in 2006) to 612,427 tonnes (in 2018) [46,65,66]. This waste was mainly produced in the Antofagasta region, and the Metropolitan region. The large amounts generated in the Antofagasta region, located in the north of Chile, can be attributed to mining activities and other associated services.

3.4. *Copper Mining Waste*

3.4.1. General Information

Chile is the world's largest producer of copper, with high-quality mineral resources located in northern Chile which has resulted in the installation of mineral processing plants over the last century (see Table 7). The copper mining industry is Chile's main economic sector, contributing to 11% of its GDP during last 10 years. Copper and its associated by-product exports represent the biggest revenue source for the country [67]. Copper production, however, produces millions of tonnes of waste; mining is generally considered as highly contaminating and unsustainable. Several measures have been implemented to optimize processes, dispose residues in an appropriate and safe manner, and to develop valorization processes.

Legislative regulations are essential to establish the responsibilities and duties of mining companies, through an exploratory and exploitative concession. Laws such as *'Closure of Mining Activities and Installations'* (Law 20.551) provide protection to the environment and to the population, as well as the mitigation of negative effects generated during the mine's lifespan and its closure [68]. Rule D28, which has been operational since December 2018, establishes the maximum permitted emissions of SO_2, arsenic, mercury and particulate matter for copper foundries. They are required to capture more than 95% of these particulate emissions [69]. The program "Alta Ley" was created by *Fundación Chile*, a non-profit organization to co-ordinate the framework of these improvements. The program developed a technological road map called 'From Copper to Innovation' which reports areas of work focused on correcting, renewing and improving the mining industry. The most important challenges include ore grades, water resources, energy resources, waste generation, gas and fine dust emissions, among others [70].

Table 7. Global copper mining (in million tonnes) in 2020; contributions from top five copper mining countries are shown [71].

Global Production	#1	#2	#3	#4	#5
	Chile	**Peru**	**China**	**D R Congo**	**U.S.A**
20	5.7	2.2	1.7	1.3	1.2

3.4.2. Typical Mining Wastes

Several types of wastes are produced during mining depending on the type of mineral exploited and its processing route. Mine waste, low grade ores and gangue represent the largest volumes of solid residues produced by the Chilean copper mining industry. Several million tonnes of soil must be removed to process valuable minerals/ores. For example, in an open pit operation, up to 500 tonnes of material may need to be removed for each tonne of copper produced. However, most of these wastes are inert materials and rocks with little environmental impact. Only a small fraction (up to 1%) may require specialist disposal. The physical stability of mining and metallurgical wastes plays a crucial role in their impact on the environmental. The chemical aspects and mineralogical composition of ores and mining waste determine the chemical stability, degradation and associated environmental impact.

A schematic representation of copper extraction routes from oxide and sulfide ores is provided in Figure 3.

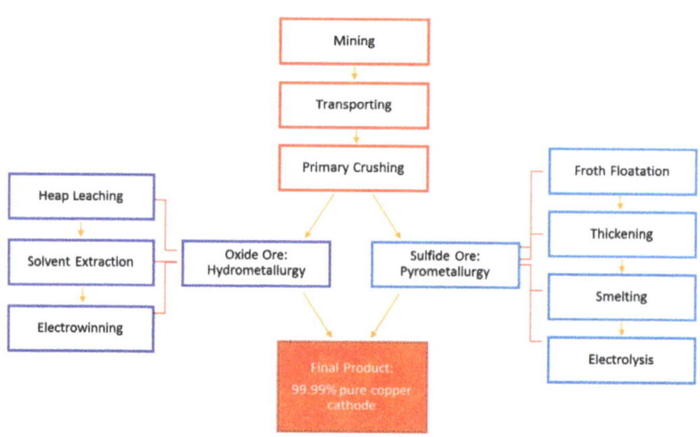

Figure 3. A schematic representation of copper extraction routes for oxide and sulfide ores [72].

A few of the focal waste sources are provided in the next section. Low grade ores are sometimes dumped separately for processing at a later stage. Metal dissolution, acid generation, bacterial leaching and effluents can lead to environmental issues at the dump sites. Froth flotation is a method that is used to concentrate copper sulfide ores by fine grinding the ore and the separation of the gangue. This process generates large amounts of tailings containing the gangue minerals and a portion of sulfides (mainly pyrite). The tailings are characterized by fine grained minerals, with a high water fraction and a high pH value. Abandoned tailing ponds constitute very serious environmental challenges. These are solid wastes produced after the flotation process, which are milled and have a fine particle size. Approximately 98% of the rock that is finely milled is disposed of, and only 2% of the rock contains copper [73]. With ore grades becoming poorer, higher volumes of rock must be removed to maintain production levels. This means that larger quantities of material will be disposed of, either as gangue or tailings. It has been estimated that copper tailings will double in size by 2035 [74]. Tailings are one of the biggest solid residues generated during the processing of copper minerals. It has been estimated that 60 to 100 tonnes of tailings are generated per 1 tonne of copper produced. The latest studies indicate that the cumulative volumes of tailings would reach ~7,900Mm^3 with a total mass of ~10,600Mt, corresponding to 742 deposits in the country. Out of 742 deposits, 104 are active, 463 inactive and 173 have been abandoned [75]. The various metallurgical wastes produced during copper processing include spent leached ore, slags, pyrometallurgical dust and gas cleaning residues. The high temperature processing of copper concentrates involves smelting, converting and refining. During these stages, slags are produced and represent the largest volumes of residues generated, with 2 to 5 tonnes of slags per 1 tonne of copper produced. Less than 1 ton of SO_2 is also formed as a gaseous biproduct [76].

3.4.3. Waste Disposal and Environmental Initiatives

As the mining industry represents the major activity in the country and generates huge amounts of residues, several measures have been taken to improve its waste and resource management. Regarding tailings generation, conventional tailings represent potential water losses and/or evaporation which is a serious issue for their appropriate disposal and management. The disposal of thickened tailings has three clear advantages, namely, good water recovery, low percolation and low fine-dust emissions. The rules pertaining to the disposal of thickened tailings have been established as per the *'Closure of Mining Activities and Installations'* law [77]. An increase in tailings generation will generate significant physical, chemical and territorial storage issues. A study in the Coquimbo region of three types of industries, i.e., mining, wastewater treatment and agriculture found that their associated wastes can cause sludges and dam sediments to encapsulate and neutralize tailings, thereby reducing the environmental impact of their storage [78].

A few accidents have occurred due to the relative instability of tailing dams. The risks are associated with the breaking off of the tailing retaining wall followed by land floods. This can occur as a result of natural events or operational design issues. In Chile, the latest risk event occurred in the 1960's when a tailing dam collapsed. It is important to assess the percolation and environmental impact of the tailing dams. These can cause harmful effects on health, water and soil. The law requires that tailings be managed in a way that they are physically and chemically stable once disposed of, and as such control is a highly important factor in preventing risks associated with natural disasters.

With regard to slags, two processing plants for slags have been built by CODELCO, one is located in El Teniente division and the other in El Salvador. These are floatation plants that recover copper and generate filtered tailings. There has also been some research on slag treatments. Investigations have confirmed the presence of Fe-C-Cu alloys and precious metals, which could be processed further to recover metals. Additionally, inert slags could be used for cement production. It has also been possible to recover iron and molybdenum from copper slags by selective reduction due to their chemical affinities [79–81]. All these byproducts are being studied and used as construction materials; chemical and physical studies have demonstrated that that have similar characteristics to the materials used in the construction industry. This is a clear example of a circular economy, as the residues produced in one industry could provide the raw material needed for a different industry, thereby enhancing the pyrometallurgical capacity in the region as well as expertise in the mining field [47].

3.5. Other Solid Wastes—Electronic Waste, Plastics

In addition to the consistent efforts to improve waste management for MSW and in the mining sector, Chile has made significant improvements in managing other wastes as well. Chile is the second biggest producer of electronic waste (e-waste) in Latin America, producing 8.9 kg of e-waste per capita per year [82]. The highest e-waste producers in Latin America are Brazil and Mexico, generating 1.5 and 1 Mtons, respectively; Chile is estimated to have produce ~200,000 tonnes of e-waste in 2018 [83]. Although this amount is much lower than the amounts produced by Brazil and Mexico, Uruguay and Chile were among highest e-waste producers per inhabitant in 2016, generating 10.8 and 8.7 kg per capita, respectively. While the typical annual growth worldwide is 3–4%, the estimated growth of e-waste in Latin America in 2018, was 5%. One of the key factors contributing to a high e-waste generation in Chile is directly related to the GDP rise during previous years. In 2012, the per capita income in Chile was 18,354 USD as compared to 1, 292 USD in Haiti. Higher income levels lead to a higher acquisition of electronic goods and the consequent generation of e-waste [84].

A typical recycling chain for e-waste consists of four steps: collection, dismantling, mechanical/physical processing and metallurgical processing [85]. Chile has developed an urban mining sector with a focus on the first two stages of the chain, i.e., collection and disassembly. The valuable components of e-waste, such as printed circuit boards (PCBs), are currently sent overseas where the material is processed. Any plastics separated from the e-waste are formally processed in the country. Some initiatives are currently being developed to implement the recovery of metals from PCBs locally. The validation of a high temperature pyrolysis process is currently being pursued through a collaboration between the Chilenter Foundation and Universidad Andrés Bello [86–88]. The purpose of this project is to concentrate valuable metals into a reduced fraction, while other components are retained as a non-metallic fraction to be used as a raw material for purposes such as carbon and energy resource. Chile has the potential of recovering valuable metals from e-waste as it has the largest pyrometallurgical capacity in the region, as well as expertise in the mining field [47].

As Chile consumes large amounts of plastic products, waste plastics are recycled at a considerable rate in the country. As a priority product considered in the *Waste Management Framework*, if packaging is created using plastic, then the extended producer responsibility plays an important role in controlling the type and quantity of plastic being used for fabrication and its subsequent disposal. Another important measure promoted by the Chilean government is to prohibit the commercial use of plastic bags in the whole of its territory, making Chile the first country in Latin America to prohibit their use [89]. Action has been taken to avoid the contamination of the land and especially the ocean across the whole country. It is important to mention that Chile has 4,300 km of coast. The Government has also launched a campaign to reduce the use of plastic straws, which has increased awareness regarding environmental impact of single-use plastics and has therefore minimized their use, either by avoiding them or by replacing them with glass or steel straws.

4. Concluding Remarks

An in-depth overview has been presented on solid waste management in relation to the global scenario, and a further special report on waste management in Chile has been presented. Given the wide variety of wastes being produced, there is no single method to manage all wastes. For example, one of the major issues in managing MSW is its collection from various households, offices and institutes, and its transportation to central processing areas. This feature represents one of the costliest aspects of waste management operations. Biodegradable organic wastes, inert wastes and the mixed nature of waste and their sorting presents additional challenges. The collection of waste is not a major issue for industrial wastes as most of it is produced locally at industrial sites. However, large volumes and relatively inert wastes, some of which may be hazardous, poses a major challenge. Through various related government projects and legislation, Chile is making very sincere multipronged efforts to manage wastes produced by its population, its industries and the mining sector. Continuous sustained efforts, an education of its populace, an awareness of waste hazards and the creation of waste management infrastructures provides the solution for sustainable waste management, the recovery of resources and a circular economy.

Author Contributions: Conceptualization, original draft: R.C. and R.K.; Supervision and resources: Y.K., I.B.; Formal analysis: J.B.K. and P.S.M. All authors have read and agreed to the published version of the manuscript.

Funding: This research received no external funding.

Institutional Review Board Statement: Not applicable.

Informed Consent Statement: Not applicable.

Data Availability Statement: Not applicable.

Conflicts of Interest: The authors declare no conflict of interest.

References

1. Global Waste Generation—Statistics & Facts. Available online: https://www.statista.com/topics/4983/waste-generation-worldwide/ (accessed on 28 August 2021).
2. What a Waste 2.0: A Global Snapshot of Solid Waste Management to 2050. Chapter 2. Available online: https://datatopics.worldbank.org/what-awaste/trends_in_solid_waste_management.html (accessed on 28 August 2021).
3. What a Waste 2.0: A Global Snapshot of Solid Waste Management to 2050. Available online: https://openknowledge.worldbank.org/handle/10986/30317 (accessed on 30 August 2021).
4. What a Waste Global Database. Available online: https://datacatalog.worldbank.org/dataset/what-waste-global-database (accessed on 30 August 2021).
5. Center for Sustainable Systems, University of Michigan. 2021. "Municipal Solid Waste Factsheet". Pub. No. CSS04-15. Available online: https://css.umich.edu/factsheets/municipal-solid-waste-factsheet (accessed on 20 August 2021).
6. Bidlingmaier, W.; Sidaine, J.M.; Papadimitriou, E.K. Separate collection and biological waste treatment in the European Community. *Rev. Environ. Sci. Bio/Technol.* **2004**, *3*, 307–320. [CrossRef]
7. What a Waste 2.0: A Global Snapshot of Solid Waste Management to 2050. Available online: https://www.worldbank.org/en/news/infographic/2018/09/20/what-a-waste-20-a-global-snapshot-of-solid-waste-management-to-2050 (accessed on 20 August 2021).
8. Patel, V.; Meka, S. Forecasting of Municipal Solid Waste Generation for Medium Scale Towns Located in the State of Gujarat, India. *Int. J. Innov. Rese. Sci. Engg. Tech.* **2013**, *2*, 4707–4716.
9. EPA. Adjusting waste generation. In *Appendix H: Methodology to Calculate Waste Generation Based on Previous Year*; United State Environmental Protection Agency: Washington, DC, USA, 1997; p. 5.
10. Rachdawong, P.; Khaodhiar, S.; Sangiampaisalsuk, N. Analysis of solid waste generation-Characteristics in Bangkok, Thailand. *J. Environ. Res.* **2000**, *22*, 80–92.
11. Sukholthaman, P.; Chanvarasuth, P.; Sharp, A. Analysis of waste generation variables and people's attitudes towards waste management system: A case of Bangkok, Thailand. *J. Mater. Cycles Waste Manag.* **2017**, *19*, 645–656. [CrossRef]
12. Liu, J.; Li, Q.; Gu, W.; Wang, C. Generation of Municipal Solid Waste in China: Evidences from Provincial Data. *Int. J. Environ. Res. Public Health* **2019**, *16*, 1717. [CrossRef]
13. Masebinu, S.O.; Akinlabi, E.T.; Muzenda, E.; Aboyade, A.O.; Mbohwa, C.; Manyuchi, M.M.; Naidoo, P. A Review on Factors affecting Municipal Solid Waste Generation. In Proceedings of the 2nd International Engineering Conference (IEC 2017) Federal University of Technology, Minna, Nigeria, 17–19 October 2017.
14. Wang, A.; Zhang, L.; Shi, Y.; Rozelle, S.; Osborn, A.; Yang, M. Rural Solid Waste Management in China: Status, Problems and Challenges. *Sustainability* **2017**, *9*, 506. [CrossRef]
15. Zeng, C.; Niu, D.; Zhao, Y. A comprehensive overview of rural solid waste management in China. *Front. Environ. Sci. Eng.* **2015**, *9*, 949–961. [CrossRef]

16. Municipal Solid Waste in the U.S.—Statistics & Facts. Available online: https://www.statista.com/topics/2707/municipal-solid-waste-in-the-united-states/ (accessed on 23 August 2021).
17. Distribution of Municipal Solid Waste Treatment and Disposal Worldwide in 2016, by Method. Available online: https://www.statista.com/statistics/916682/global-municipal-solid-waste-treatment-by-method/ (accessed on 22 August 2021).
18. EPA Victoria. Industrial Waste Resource Guidelines. Solid Industrial Waste Hazard Categorisation and Management. Available online: https://www.epa.vic.gov.au/about-epa/publications/iwrg631 (accessed on 23 August 2021).
19. Maczulak, A.E. *Pollution: Treating Environmental Toxins*; Infobase Publishing: New York, NY, USA, 2010; p. 120. ISBN 9781438126333.
20. United States Environmental Protection Agency EPA. Facts and Figures about Materials, Waste and Recycling. Available online: https://www.epa.gov/facts-and-figures-about-materials-waste-and-recycling/national-overview-facts-and-figures-materials (accessed on 21 August 2021).
21. The Role of Steel Manufacturing in the Global Economy. 2019. Available online: https://www.worldsteel.org/en/dam/2520Summary.pdf (accessed on 10 August 2021).
22. Steel Industry Co-Products. Available online: https://www.worldsteel.org/publications/position-papers/co-product-position-paper.html (accessed on 21 August 2021).
23. Solid Waste Management in a Steel Plant. Available online: https://www.ispatguru.com/solid-waste-management-in-a-steel-plant/ (accessed on 23 August 2021).
24. U.S. Department of Transportation. User Guidelines for Waste and Byproduct Materials in Pavement Construction. Available online: https://www.fhwa.dot.gov/publications/research/infrastructure/structures/97148/ssa1.cfm (accessed on 23 August 2021).
25. Kongkarat, S.; Khanna, R.; Koshy, P.; O'Kane, P.; Sahajwalla, V. Use of waste Bakelite as a raw material resource for recarburization in steelmaking processes. *Steel Res. Int.* **2011**, *82*, 1228–1239. [CrossRef]
26. *The Management of Steel Industry By-Products and Waste ed*; International Iron and Steel Institute: Brussels, Belgium, 1987.
27. Rahman, M.; Khanna, R.; Sahajwalla, V.; O'Kane, P. The influence of ash impurities on interfacial reactions between carbonaceous materials and EAF slag at 1550 C. *ISIJ Int.* **2009**, *49*, 329–336. [CrossRef]
28. Aluminium—The Second Most Used Metal in the World. Available online: https://etem.com/blog/aluminium-the-second-most-used-metal-in-the-world (accessed on 3 September 2021).
29. Countries with the Largest Smelter Production of Aluminium from 2016 to 2020. Available online: https://www.statista.com/statistics/264624/global-production-of-aluminum-by-country/ (accessed on 3 September 2021).
30. Zhang, R.; Zheng, S.; Ma, S.; Zhang, Y. Recovery of alumina and alkali in Bayer red mud by the formation of andradite-grossular hydrogarnet in hydrothermal process. *J. Hazard. Mater.* **2011**, *189*, 827–835. [CrossRef]
31. Khanna, R.; Konyukhov, Y.V.; Ikram-ul-haq, M.; Burmistov, I.; Cayumil, R.; Belov, V.A.; Rogachev, S.O.; Leybo, D.V.; Mukherjee, P.S. An innovative route for valorising iron and aluminium oxide rich industrial wastes: Recovery of multiple metals. *J. Environ. Manag.* **2021**, *295*, 113035. [CrossRef]
32. Patel, S.; Pal, B.K. Current Status of an Industrial Waste: Red Mud an Overview. *IJLTEMAS* **2015**, *7*, 1–16.
33. Yolandi, S.; Oberholster, S.P.; Somerset, V. A decision-support framework for industrial waste management in the iron and steel industry: A case study in Southern Africa. *Case Stud. Chem. Environ. Eng.* **2021**, *3*, 100097.
34. Khairul, M.A.; Zanganeh, J.; Moghtaderi, B. The composition, recycling and utilisation of Bayer red mud. *Resour. Conserv. Recyc.* **2019**, *141*, 483–498. [CrossRef]
35. Totten, G.E.; MacKenzie, D.S. *Handbook of Aluminium: Volume 2: Alloy Production and Materials Manufacturing*; Marcel Dekker, Inc.: New York, NY, USA, 2003; Volume 2, ISBN 0-8247-0896-2.
36. The Aluminium Association: Primary Production. Available online: https://www.aluminum.org/industries/production/primary-production (accessed on 7 July 2021).
37. OECD Global Forum on Environment. Focusing on Sustainable Materials Management. Materials case study 2: Aluminium. Available online: https://www.oecd.org/env/waste/46194971.pdf (accessed on 23 August 2021).
38. United States Environmental Protection Agency EPA. Bauxite and Alumina Production Wastes. Available online: https://www.epa.gov/radiation/tenorm-bauxite-and-alumina-production-wastes (accessed on 7 July 2021).
39. The Aluminium Association: Recycling. Available online: https://www.aluminum.org/industries/production/recycling (accessed on 7 July 2021).
40. Liu, Y.; Lin, C.; Wu, Y. Characterization of red mud derived from a combined Bayer process and bauxite calcination method. *J. Hazard. Mater.* **2007**, *146*, 255–261. [CrossRef]
41. Mayes, W.M.; Jarvis, A.P.; Burke, I.T.; Walton, M.; Feigl, V.R.; Klebercz, O.; Gruiz, K. Dispersal and attenuation of trace contaminants downstream of the Ajka bauxite residue (red mud) depository failure. Hungary. *Environ. Sci. Technol.* **2011**, *45*, 5147–5155. [CrossRef]
42. Wang, J.; Zhao, P. Method of De-Alkalizing Red Mud and Recovering Aluminium and Iron. Available online: https://patents.google.com/patent/CN103031443A/en (accessed on 7 July 2021).
43. Kumar, R.A.S.; Premchand, J.P. Utilization of Iron Values of Red Mud for Metallurgical Applications. *Environ. Waste Manag.* **1998**, 108–119. [CrossRef]
44. Gronostajski, J.; Marciniak, H.; Matuszak, A. New methods of aluminium and aluminium-alloy chips recycling. *J. Mater. Process. Technol.* **2000**, *106*, 34–37. [CrossRef]

45. The SPL Waste Management Challenge in Primary Aluminum. Available online: https://www.lightmetalage.com/news/industry-news/smelting/the-spl-waste-management-challenge-in-primary-aluminum/ (accessed on 7 July 2021).
46. Ministerio del Medio Ambiente. Informe del Estado del Medio Ambiente (IEMA) 2020, Chapter 10, p. 2. "Residuos"—SINIA. Available online: https://sinia.mma.gob.cl/index.php/resiudos/ (accessed on 29 August 2021).
47. Ulloa, G.; Cayumil, R.; Sánchez, M. Urban Mining in Chile: State of the art. In Proceedings of the Fourth Symposium of Urban Mining and Circular Economy, Bergamo, Italy, 21–22 May 2018.
48. Basel Convention. *On the Control of Transboundary Movements of Hazardous Wastes and Their Disposal*; Basel Convention: Geneva, Switzerland, 2014.
49. United Nations Environment Program. Stockholm Convention on Persistent Organic Pollutants. United Nations, Geneva. Nations, Geneva. 2019. Available online: https://www.unep.org/annualreport/2019/index.php (accessed on 30 August 2021).
50. CONAMA. Plan Nacional de Implementación para la Gestión de los Contaminantes Orgánicos Persistentes (COPs) en Chile. 2005, 1–67. Available online: https://www.informea.org/en/plan-nacional-de-implementaci%C3%B3n-para-la-gesti%C3%B3n-de-los-contaminantes-org%C3%A1nicos-persistentes-cops-en (accessed on 29 August 2021).
51. MERCOSUR: A Status Report and Prospects for Canada-Mercosur Relations. Available online: https://www.sice.oas.org/TPD/CAN_MER/Studies/MercosurFocal_e.pdf (accessed on 21 August 2021).
52. Ministerio del Medio Ambiente. Santiago Recicla. Available online: http://centralenergia.cl/uploads/2011/09/Bases_generales_del_medio_ambiente_LEY-19300.pdf (accessed on 20 August 2021).
53. Available online: https://www.informea.org/en/framework-law-waste-management-law-no-20920-2016 (accessed on 10 August 2021).
54. Available online: https://eticayseguridad.uc.cl/documentos/comite-seguridad/normativa-seguridad/127-dcto-148-residuos-peligrosos/file.html (accessed on 21 August 2021).
55. Ministerio del Medio Ambiente. Santiago Recicla. Available online: http://www.santiagorecicla.cl/ (accessed on 30 May 2021).
56. Glavič, P.; Lukman, R. Review of sustainability terms and their definitions. *J. Cleaner Product.* **2007**, *15*, 1875–1885. [CrossRef]
57. Ministerio del Medio Ambiente. Economía Circular. Available online: https://mma.gob.cl/economia-circular/ (accessed on 1 September 2021).
58. Average per Capita Generation of Municipal Solid Waste Worldwide in 2016, by Region. Available online: https://www.statista.com/statistics/916618/global-per-capita-generation-of-municipal-solid-waste-by-region/ (accessed on 1 September 2021).
59. Blazquez, C.; Paredes-Belmar, G. Network design of a household waste collection system: A case study of the commune of Renca in Santiago, Chile. *Waste Manag.* **2020**, *116*, 179–189. [CrossRef]
60. Pontificia Universidad Católica. *Estudio Caracterización de Residuos Sólidos Domiciliarios en la Region Metropolitana*; UCV: Valparaíso, Chile, 2006.
61. Chile and Canada Partner to Reduce Emissions from the Waste Management Sector. Available online: https://www.ccacoalition.org/en/news/chile-and-canada-partner-reduce-emissions-waste-management-sector (accessed on 23 August 2021).
62. Chile–Developing a Legal Framework for EPR in Chile. Available online: https://prevent-waste.net/wp-content/uploads/2020/09/Chile.pdf (accessed on 28 August 2021).
63. Rojas, A.; Yabar, H.; Mizunoya, T.; Higano, Y. The Potential Benefits of Introducing Informal Recyclers and Organic Waste Recovery to a Current Waste Management System: The Case Study of Santiago de Chile. *Resources* **2018**, *7*, 18. [CrossRef]
64. AVINA. *Politicas Publicas Para la Inclusion de los Recicladores de Base al Sistema de Gestion de Residuos Municipales en Chile*; AVINA: Santiago, Chile, 2013. (In Galician)
65. Hutchison, I.P.G.; Ellison, R.D. Mine Waste Management: A Resource for Mining Industry Professionals, Regulators and Consulting Engineers. United States, 1992. Available online: https://www.osti.gov/biblio/6715101-mine-waste-management-resource-mining-industry-professionals-regulators-consulting-engineers (accessed on 10 August 2021).
66. Ministerio de Energía. Energías Renovables. Available online: http://www.energia.gob.cl/energias-renovables (accessed on 31 May 2021).
67. Sociedad Nacional de Minería. Fundamentos y Desafíos para el Desarrollo Minero. 2017. Available online: https://www.sonami.cl/v2/centro-de-documentacion/publicaciones-tecnicas/ (accessed on 31 May 2021).
68. SERNAGEOMIN, n.d. Cierre de Faenas Mineras. Available online: https://www.sernageomin.cl/cierre-de-faenas-mineras/ (accessed on 31 May 2021).
69. Ministerio Del Medio Ambiente, 2018c. Norma de Emisión Para Fundiciones de Cobre y Fuentes Emisoras de Arsénico. Available online: https://www.leychile.cl/Navegar?idNorma=1057059 (accessed on 11 August 2021).
70. Fundación Chile. Desde el Cobre a la Innovación. Roadmap Tecnológico 2015–2035. Available online: http://programaaltaley.cl/wp-content/uploads/2016/04/Roadmap-Tecnologico-Alta-ley.pdf (accessed on 12 August 2021).
71. Mine Waste Management: A Resource for Mining Industry Professionals, Regulators and Consulting Engineers. Available online: https://www.nsenergybusiness.com/news/top-five-copper-mining-countries/ (accessed on 6 July 2021).
72. Copper Mining and Processing: Processing Copper Ores. Available online: https://superfund.arizona.edu/resources/learning-modules-english/copper-mining-and-processing/processing-copper-oreshttps://www.mwen.info/docs/imwa_1999/IMWA1999_Wiertz_403.pdf (accessed on 6 July 2021).

73. Schlesinger, M.; King, M.; Sole, K.; Davenport, W. Extractive Metallurgy of Copper (Fifth). Sociedad Nacional de Minería, 2017. Fundamentos y desafíos para el desarrollo minero. Available online: https://www.elsevier.com/books/extractive-metallurgy-of-copper/schlesinger/978-0-08-096789-9 (accessed on 6 July 2021).
74. Roadmap Tecnológico de la Minería. Available online: https://corporacionaltaley.cl/?page_id=952&lang=en (accessed on 6 July 2021).
75. SERNAGEOMIN, 2018. Preguntas Frecuentes sobre Relaves. Available online: https://www.sernageomin.cl/preguntas-frecuentes-sobre-relaves/ (accessed on 6 July 2021).
76. Moskalyk, R.R.; Alfantazi, A. Review of copper pyrometallurgical practice: Today and tomorrow. *Miner. Eng.* **2003**, *17*, 893–919. [CrossRef]
77. Minería Chilena, 2009. Relaves Espesados a la Vanguardia. Available online: https://www.mch.cl/reportajes/relaves-espesados-a-la-vanguardia/ (accessed on 6 July 2021).
78. El Ovallino. Tecnosuelos: La Alternativa que Busca "Reverdecer" los Relaves Mineros. 2019. Available online: http://www.elovallino.cl/economia/tecnosuelos-alternativa-que-busca-reverdecer-relaves-mineros (accessed on 6 July 2021).
79. Zongjie, L.; Junrui, C.; Zengguang, X.; Yuan, Q.; Jing, C. A Comprehensive Review on Reasons for Tailings Dam Failures Based on Case History. *Adv. Civil Eng.* **2019**, *2019*, 4159306.
80. Palacios, J.; Sánchez, M. Wastes as resources: Update on recovery of valuable metals from copper slags. *Miner. Process. Extract. Metall.* **2011**, *120*, 218–223. [CrossRef]
81. Sánchez, M.; Sudbury, M. Reutilisation of primary metallurgical wastes: Copper slag as a source of copper, molybdenum and iron–brief review of test work and the proposed way forward. In Proceedings of the Third International Slag Valorisation Symposium, Leuven, Belgium, 19–20 March 2013.
82. Forti, V.; Baldé, C.P.; Kuehr, R.; Bel, G. The Global E-waste Monitor 2020: Quantities, flows and the circular economy potential. United Nations University (UNU)/United Nations Institute for Training and Research (UNITAR)—Co-Hosted SCYCLE Programme, International Telecommunication Union (ITU) & International Solid Waste Association (ISWA), Bonn/Geneva/Rotterdam. Available online: http://ewastemonitor.info/ (accessed on 10 September 2021).
83. Magalini, F.; Khetriwal, D.S.; Huisman, J.; Nanorom, I.C. Electronic Waste (E-Waste) Impacts and Mitigation Options in the Off-grid Renewable Energy Sector. 2016. Available online: https://assets.publishing.service.gov.uk/media/58482b3eed915d0b12000059/EoD_Report__20160825_E-Waste_Study_Final-31.08.16.pdf (accessed on 10 September 2021).
84. Solving the E-Waste Problem. Available online: https://www.step-initiative.org/ (accessed on 10 September 2021).
85. Tsydenova, O.; Bengtsson, M. Chemical hazards associated with treatment of waste electrical and electronic equipment. *Waste Manag.* **2011**, *31*, 45–58. [CrossRef]
86. Cayumil, R.; Khanna, R.; Ikram-Ul-Haq, M.; Rajarao, R.; Hill, A.; Sahajwalla, V. Generation of copper rich metallic phases from waste printed circuit boards. *Waste Manag.* **2014**, *34*, 1783–1792. [CrossRef]
87. Cayumil, R.; Khanna, R.; Rajarao, R.; Mukherjee, P.S.; Sahajwalla, V. Concentration of precious metals during their recovery from electronic waste. *Waste Manag.* **2016**, *57*, 121–130. [CrossRef]
88. Cayumil, R.; Ikram-Ul-Haq, M.; Khanna, R.; Saini, R.; Mukherjee, P.S.; Mishra, B.K.; Sahajwalla, V. High temperature investigations on optimising the recovery of copper from waste printed circuit boards. *Waste Manag.* **2018**, *73*, 556–565. [CrossRef]
89. Ministerio del Medio Ambiente. Chao Bolsas Plásticas. Available online: http://chaobolsasplasticas.cl/ (accessed on 6 July 2021).

sustainability

MDPI

Article

Development of a Software System for Selecting Steam Power Plant to Convert Municipal Solid Waste to Energy

Rotimi A. Ibikunle [1,*], Isaac F. Titiladunayo [2] and Basil O. Akinnuli [3]

1 Department of Mechanical Engineering, College of Engineering, Landmark University, PMB 1001, Ipetu Road, Omu-Aran 251103, Nigeria

2 Department of Mechanical Engineering, School of Engineering, Federal University of Technology, PMB 704, Akure 340106, Nigeria; ftitiladunayo@yahoo.com

3 Department of Industrial and Production Engineering, School of Engineering, Federal University of Technology, PMB 704, Akure 340106, Nigeria; ifembola@yahoo.com

* Correspondence: ibikunle.rotimi@lmu.edu.ng

Abstract: A software system that enhances the selection of appropriate power plant capacity that will convert combustible municipal solid waste (MSW) into energy was developed. The aggregate of waste to be converted was determined and the corresponding heating value was established. The capacities of steam power plants' components required for the conversion were determined, using thermodynamic mathematical models. An algorithm based on models used to determine the energy potential, the power potential of MSW, the capacities of the components of the steam power plant, were translated into computer soft code using Java programming language; saturated steam and superheated steam tables, together with the thermodynamic properties of the power plant required were incorporated into the soft code. About 584 tons of MSW having a heating value of 20 MJ/kg was the quantity of waste experimented for energy generation. This information was input into the software as data and was processed. Then, the software was able to predict 3245.54 MWh energy potential for the quantity of waste, and electrical power potential of 40.54 MW. The capacities of the steam power plant components that were predicted include 100.35 MW of boiler power, 40.54 MW of turbine power, and 59.80 MW of condenser power. The methodology adopted will make it easy for the managers in the waste-to-energy sector to appropriately select the suitable capacity of the required steam power plant that can convert any quantify of MSW at any geographical location, without going through the engineering calculation and stress or rigor involved in the plant capacity design. Moreover, the accuracy obtained for the software is greater than 99%.

Keywords: decision support system; combustible MSW; energy potential; power potential; selection of power plant

Citation: Ibikunle, R.A.; Titiladunayo, I.F.; Akinnuli, B.O. Development of a Software System for Selecting Steam Power Plant to Convert Municipal Solid Waste to Energy. *Sustainability* **2021**, *13*, 11665. https://doi.org/10.3390/su132111665

Academic Editors: Rita Khanna, Yury Konyukhov and Igor Burmistrov

Received: 2 August 2021
Accepted: 31 August 2021
Published: 21 October 2021

Publisher's Note: MDPI stays neutral with regard to jurisdictional claims in published maps and institutional affiliations.

1. Introduction

Municipal solid waste (MSW), is a collection of materials resulted from man's activities with his environment on daily basis, which are considered as not useful and are disposed of as wastes. MSW generation is on the high side over the years, around the globe, due to the increase in population, changes in taste, fashions, advancement in technology, and consumerism growth [1,2]. Islam [3] and Ibikunle [4,5] reported that the growth in MSW generation is due to the global quest for urbanism, social, and industrial development. Johari [6] stated that the sporadic increase in a waste generation will necessitate environmental challenges unless it is adequately and promptly managed. Ibikunle [7] remarked that the waste generated in the Ilorin metropolis of Kwara State is a huge one, and the waste management system available is insufficient; this makes people indulge in indiscriminate disposal of wastes which results in pollution, an unsightly scene, and blockage of waterways.

The global record on MSW production showed a generation rate of 680 million tons/annum in the year 2000 and it increased to 1300 million tons/annum in the year 2010. A projection of 2200 million tons/annum was made for the year 2025 while 4200 million tons/annum was projected for the year 2055 [8]. The rate of waste generation (MSW/capita/day) is a principal environmental pressure indicator required to estimate the intensity of waste generation for effective management planning [7]. Moreover, it is also useful in comparison of the degree of MSW generation between a nation and another. In the United Kingdom, the MSW rate of generation was recorded to be 1.34 kg/capita/day, 2.13 kg/capita/day for the United States, 2.00 kg/capita/day for South Africa, 0.09 kg/capita/day for Ghana, and 0.58 kg/capita/day for Nigeria, according to Parashar [9]. Nevertheless, Ibikunle [7] reported the rate of MSW production in the Ilorin metropolis to be 0.78 kg/capita/day, which is 20% greater than the rate estimated for waste production in Nigeria in 2016. The rate of growth in waste production in the nations is quite alarming, and this poses a threat for the waste management system especially in the Ilorin metropolis [7,10].

Lately, there is a growth in the power generation (technology) sector, utilizing renewable resources as an energy basis. Despite the development and expansion in power technology by utilizing renewable resources, 41% of Nigeria still does not have access to electricity [11]. This means only about 115 million Nigerians have access to electricity, out of about 196 million. Wind, hydro, solar, and biomass are the most utilized renewable energy resources for energy production throughout the globe, since the drive for sustainable and renewable energy [12]. It was reported by Ogunjuyigbe [13] and Ibikunle [4] that recovery of clean energy from MSW will promote a clean and green living environment for the growth of a nation and the standard of living of her citizens.

The governments of many nations are strategizing to have the rate of renewable energy (RE) generation increase by reducing costs while ameliorating the rate of greenhouse gases (GHGs) emission and fossil fuels dependence. This will ensure a clean and green environment, clean energy, and green industry, and the availability of more job opportunities and the like [14]. Fundamentally, energy is required as a strategic resource in the foundation formation for social and economic development. However, the rapid demographic growth and socio-economic development require high demand for energy [15]. It is predicted that the global demand for primary energy in 2050 will increase by 80%, which will pose a threat to energy security [16]. There is quite a large demand for power that is characterized by supply that does not satisfy the domestic and commercial demand. The need to secure new renewable energy and power sources is a principal global concern because of the constant depletion of fossil fuels. Moreover, it is a compulsory task now to ameliorate the emission of carbon dioxide (CO_2) the prominent greenhouse gas (GHG) because of its consequential effect on climate change [2,17].

However, in urban cities of both developing and underdeveloped nations, there is always a massive generation of MSW, which is associated with the standard of living in metropolitan cities. Efficient management of MSW in most cities of Nigeria is a big issue [18,19]. In May (2019), the worldometer [20] reported the population of China to be 1418 million people, US—328 million, Japan—126 million, Germany—82 million, UK—66 million, and France—65 million people. The MSW generation in China is 46 million tons/year, in the US it is 262.4 million tons/year, Japan is 44 million tons/year, Germany is 51 million tons/year, the UK is 32 million tons/year, and France is 33 million tons/year [21]. Worldometer in May 2019 reported the population density of Nigeria, Ethiopia, Egypt, Tanzania, and South Africa to be 196 million, 108 million, 99 million, 59 million, and 57 million respectively. According to Kosuke [22], South Africa has a generation rate of 2.0 kg/capita/day, Nigeria has 0.58 kg/capita/day, Ethiopia has 0.3 kg/capita/day and Tanzania has 0.26 kg/capita/day. Therefore, by appropriation based on the generation rate and the population density, the MSW generated in Nigeria was estimated to be 42 million tons/year, Ethiopia was 12 million tons/year, Egypt was 44 million tons/year, Tanzania was 6 million tons/year, and South Africa was 42 million tons/year.

The number of power plants in China, US, Japan, Germany, the UK, and France is 166, 88, 822, 78, 14, and 37 respectively; while power generation in China is 18.7 million MWh, US is 14 million MW, Japan is 1.9 million kWh, Germany is 5768 GWh, the UK is 2782 GWh and 1999 GWh is in France [23]. Despite the enormous waste generated in African nations, and the recent development in WTE technology, Africa is still behind in waste-to-energy (WTE) practices. The first waste-to-energy (WTE) power plant in Africa is 110 MW of Ethiopia, and Ghana is proposing a 60 MW Armech thermal power plant [24]. Oladende [25] reported that the Nigeria Electricity Supply Industry (NESI) has 12 power stations in Nigeria with a total power output of 11,165 MW and yet could not produce power during an off-peak period of the Christmas holiday; moreover, the only 26,000-L capacity biogas power plant, at Ikosi-Ketu market in Lagos, with about 10 KVA daily capacity, has not commenced operation. In South Africa, the renewable energy project put in place by the renewable independent power producer procurement programme (REIPPPP) is about 6327 MW altogether, in which landfill and biogas take 60 MW altogether [26]. The Ilorin metropolis that is selected for this study produces 302,000 tons of MSW per year, with a generating rate of 0.78 kg/capita/day; 70% of the aggregate waste generated is combustible, yet the city still faces an energy crisis [7]. The power supply by the Power Holding Company of Nigeria (PHCN) to the Ilorin metropolis is insufficient for her social and economic demand. Therefore, if energy recovery from waste components is adopted in Ilorin, it will provide a dual solution viz., efficient waste management and provision of alternative energy to energy from fossil fuels, which can complement the power supply from PHCN.

To practice WTE in any environment, the sufficiency and efficiency of the waste fractions required must be ascertained; the sufficiency of MSW is about the quantity of the waste fractions available for energy production and the regular flow in terms of the generation rate. The efficiency of waste is about the heating value and the correlation between the physicochemical properties of the MSW and its calorific value. When the sufficiency and the efficiency of the waste have been determined, then the choice of the appropriate WTE technology can be made. In this study, software was developed using Java programming language, an algorithm based on equations used to determine energy potential and electrical power potential of the municipal solid waste. The mathematical models used to determine the capacity of each of the equipment in the waste-fired power plant was prepared, which is translated into computer soft code using Java programming language, which is one of the common programming languages used in numerical analysis as well as a unified modeling language (UML) [27].

This study aims to provide software that will serve as a management and technical decision tool, to predict the capacity of steam power equipment that will convert MSW into energy at any geographical location, provided the available quantity of waste for energy and the low heating value (LHV) is also established. This study solves one of the technical challenges that could be encountered while trying to ascertain the capacity of the plant required. Nevertheless, the software developed in this study will predict the available energy and power potential in the waste to be converted, as well as the thermodynamic properties of the stages involved in the process. This software is referred to as a power plant selection support system because it helps in taking decisions on the appropriate capacity of the steam power plant required for a specific quantity of waste when its low heating value has been ascertained. The accuracy of the software is established by comparing the calculated values to the values generated by the software, and it is found to be above 99%.

Decision Support Systems (DSSs)

The computer-based interactive systems adopted to help the users to make the right decision or choice of activities and judgment are called DSSs; this is characterized by data bank and retrieval device. It enhances access to information and restoration of functions that support modeling and reasoning based on models. DSSs also support solving problems and framing. In most situations, the standard or quality of decisions taken can be used to correct the human error involved during decision-making. Disciplines that include operations research, economics, and statistics develop various methods for decision making. The methods can be enhanced by different scientific techniques that are based on artificial intelligence. Computer programs that either operate as tools on their own or as integrated computing environments are adopted for complex decisions. Such computing environments are called decision support systems (DSSs). The broad concept of a DSS is responsible for the variation in its definitions, therefore, every definition now depends on the author's point of view. To embrace all perspectives, a DSS can be defined as a computer-based system that aids users in judgment and choice of activities [28]. This is the reason the software developed in this study, to select the appropriate capacity of steam power plant that will convert any quantity of combustible MSW to energy in any geographical location, is considered as a decision support system.

The design of a DSS and its analysis is so complex that it involves the usage of specific and adequate instruments and methodologies to model the decision processes. Customized and developed features can be used to implement a DSS for decision activities or by utilizing a generalized DSS that can later be customized. DSSs can either be specifically developed for an establishment, using programming languages or using DSS generators. The strategy involved in the software development either engages general-purpose programming language (GPL), such as C++, PASCAL, BASIC, or COBOL, or engages the use of fourth-generation language (4GL), such as VISUAL BASIC .NET, C# .NET, VISUAL J# .NET, DELPHI, JAVA, or VISUAL C++ [29]. The most appropriate area for the installation of a biomass power plant can best be selected by managers using a spatial decision support system. Fuzzy logic and fuzzy membership functions are used for the creation of criteria layers and suitability maps. Multicriteria Decision Analysis methodology (Analytical Hierarchy Process) combined with fuzzy system elements for the determination of the weight coefficients of the participating criteria are used [30]. Fatima [31] incorporated a decision support system in a power plant simulation tool to provide the qualitative synthesis needed for the plant design process, and to assist the design engineers in performing a better choice-evaluation for technical quantitative and qualitative analysis.

Konstantinos [32] stated that decision support systems (DSS) are an approach designed as a tool to help managers take decisions by accelerating the relevant processes. The DSS developed in this study is designed to accelerate processes required in the design of power plants, and concurrently predict the energy and power potentials available in a quantity of MSW. This DSS can predict the energy and power potentials of the MSW fractions that are available for power generation, predict the thermodynamic properties involved in each stage of the power cycle of the steam power plant, and also predict the capacity of each of the equipment in the power plant that will convert a particular quantity of MSW whose net heating value is known.

2. Materials and Methods

The MSW materials investigated in this article are the 9 combustible MSW fractions of the total 19 waste fractions identified in the Ilorin metropolis. These include paper, packaging boxes, grass/garden trimmings, rags, wood, food residue, nylon, polypropylene sacks, and plastic bottles. The quantity of the MSW available per day for energy production was established, and the net heating value was determined using an *e-2k* combustion calorimeter. The heat energy and the power potentials of the MSW were determined. Rankine reheat cycle was used to design the capacities of the steam power plant equipment required for energy production. The mathematical models utilized in determining the energy potential and the designing of the power plant capacity are used to form the algorithm used in the development of the software, called the decision support system. The software interfaces were developed using Java swing utilities that comprise JFrames, JTextrafields, JButtons, and JEditorpane for data display. Action listeners and action events were used to control the button functionalities, handle calculations, and panel/Editorpane displays [25].

2.1. Determination of the Aggregate MSW Generated in Ilorin/Year, Using Collection Facts and Figures

The aggregate MSW produced in the Ilorin metropolis of Kwara State between January and June 2021 was estimated by using the mathematical model in Equation (1), without the aid of a weighing bridge, as suggested by Kosuke [22] and Ibikunle [4,5,10].

$$MSW_{gen.} = \sum_{j=1}^{365} \sum_{i=1}^{n} (C_i \times V_i \times d_i \times t_{ij}) \times (0.74) \tag{1}$$

The total number of trucks is n, the capacity of a truck is C_i (m^3/truck), the loading volume ratio of a truck is V_i, the density of MSW loaded on the truck is d_i (tons/m^3), and the number of trips by truck is t_{ij} on day j (frequency of trips/day).

2.2. Characterization of the MSW Fractions

It was reported by NT ENVIR 001 [33], Issam [34], and Ibikunle [4,5,10] that the huge MSW collected from different locations in the city can be effectively characterized through random sampling of some selected heaps of wastes deposited in the dumpsite. NT ENVIR 001 [33] suggested that 240 L of a bin full of MSW are appropriate as a unit sample. The waste streams were characterized twice a week for six months, thereby making 48 different batches of samples for the characterization investigation at Lasoju dumpsite to avoid any consequences from the effect of insufficient samples. Each component identified was sorted into different receptacles and weighed.

2.3. Physicochemical and Heating Value Analysis of MSW

2.3.1. Moisture Content Analysis of MSW Components

The moisture content of the MSW combustible fractions was estimated using an electric oven (model DHG 9053) with a 200 °C capacity. About 1 g of powder of the sample of each component was measured into a crucible and dried in the oven, maintained at a temperature of 110 °C for an hour. The weight loss is considered as the moisture content of the waste component as suggested by Vairam [35], Shi [36], Titiladunayo [7], and Ibikunle [4,7,37] using Equation (2).

$$MC = \frac{W_2 - W_3}{W_2 - W_1} \times 100 \tag{2}$$

where MC is the percentage moisture content, W_1 is the mass of empty crucible (g), W_2 is the mass of crucible and sample (g), W_3 is the mass of crucible and sample after heating (g), $W_2 - W_1$ is the mass of sample (g), and $W_2 - W_3$ is the moisture content (g).

2.3.2. Ultimate Analysis of MSW Components

Ultimate analysis was investigated to ascertain the quantitative value of the carbon (C), hydrogen (H), nitrogen (N), sulphur (S), and oxygen (O) percentage content that is available in the MSW components. The investigation was carried out using an elemental analyzer (Flash EA 1112 model), based on ASTM D5291. About 0.5 g of powdered sample of each component was measured into a crucible and combusted. The oxides of carbon, hydrogen, nitrogen, and sulphur produced were analyzed by thermal conductivity detector (TCD), and the electrical signal emerged was processed by 'Eager 300 software' to give the percentages of elements contained in the sample. The samples were replicated, and an average of the values obtained is considered as the typical value [2].

2.4. Estimation of the Heating Value of the MSW Combustible Fractions

Islam [3] and Shi [33] suggested that the high heating value (HHV) of solid fuel can be determined by using *e-2k* combustion calorimeter, shown in Figure 1. The low high heating value (LHV) of the nine combustible waste fractions presented in Table 3 was obtained by applying Equation (3), as adopted by Islam [3], Ibikunle [4], Ibikunle [10], Ibikunle [37], and Kumar [38], as well as adopting Dulong and Steuer models in Equations (4) and (5) respectively. The average of the LHV obtained from the three equations is considered the typical LHV [4,7].

$$LHV^{i} = \sum_{1}^{9} W_{\text{msw}}(\%) \times HHV_{\text{msw}} \quad (3)$$

where W_{msw} is the weight fraction (%) of the combustible MSW and HHV_{msw} is the high heating value obtained by using a bomb calorimeter [4,10,37–39].

$$LHV^{ii} = 81C + 342.5\left(H - \frac{O}{8}\right) + 22.5S - 6(W + 9H) \quad (4)$$

$$LHV^{iii} = 81\left(C - \frac{3}{8}O\right) + 57\frac{3}{8}O + 345\left(H - \frac{O}{16}\right) + 25S - 6(W + 9H) \quad (5)$$

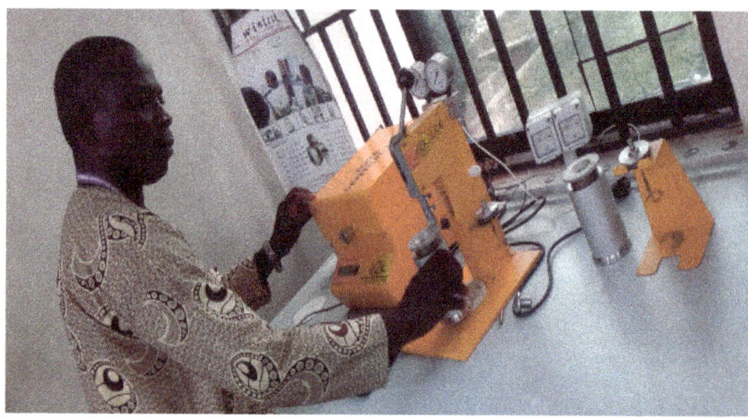

Figure 1. *e-2k* Combustion for HHV determination [27].

In Equations (4) and (5), LHV^{ii} and LHV^{iii} are the heating values estimated by adopting Dulong and Steuer models respectively, and carbon (C), hydrogen (H), sulfur (S), and oxygen (O) are the chemical elements obtained from the ultimate analysis.

2.5. Determination of the Energy Potential and Electrical Power Potential of the MSW

2.5.1. Determination of Energy Potentials (EP_{msw}) of the Municipal Solid Waste

Daura [39] and Ibikunle [10] reported that the energy potential (EP_{msw}) of MSW could be determined by applying Equation (6):

$$EP_{msw} = LHV_{msw} \times w_{msw} \times \frac{1000}{3.6}\ (\text{kWh}) \quad (6)$$

where EP_{msw} is the energy potential from MSW, W_{msw} (tons) is the weight of MSW, LHV_{msw} is the net low heating value of the MSW (MJ/kg). Conversion ratio (1 kWh = 3.6 MJ).

$$EP_{msw} = 20 \times 584 \times 227.7 = 3,244,444 \text{ kWh or } 3.2 \text{ GWh}$$

2.5.2. Determination of the Electrical Power Potential (EPP_{msw}) of the Municipal Solid Waste

The Electrical Power Potential (EPP_{msw}) of the MSW Is Determined Using Equation (7) as Suggested by Ibikunle [7,10] and Daura [39]

$$EPP_{msw} = 277.8 \times LHV_{msw} \times \frac{w_{msw}}{24} \times \eta \text{ (kW)} \tag{7}$$

$$EPP_{msw} = 40,602 \text{ kW}$$

where η is the conversion efficiency in a power plant, which is within a range of 20–40% as reported by Muhammad [40] and Ibikunle [7,10]. However, conversion efficiency of 30% is adopted for this work.

$$\text{Power to grid } (GP) = EPP_{msw} \times \eta_g \times \eta_p \times \frac{1}{1000} \text{ (MW)} \tag{8}$$

$$GP = 27.3 \text{ MW}$$

where η_g is the generator efficiency (selected is 90%), η_p is the transmission efficiency (selected is 75%) of turbine work (W_T). The generator efficiency range is 85–90% and turbine efficiency is within a range of 75–80% [41].

2.6. *Determination of the Steam Power Plant Capacity Required for Energy Production*

Thermodynamic analysis of Rankine cycle is used to enhance the reliability and efficiency of steam power plants [42]. In this study, the reheat Rankine cycle presented in Figure 2b is used to determine the capacity of the steam power plant that will convert the waste to electrical power. Rankine reheat cycle in Figure 2b was used for the estimation of the steam power plant capacity because it has a higher cycle efficiency compared to the ordinary superheat Rankine cycle shown in Figure 2a. The combination of temperature (400 °C) and pressure (40 bar) is commonly used in the capacity determination/design for steam power plants to minimize investment costs. Temperature higher than 400 °C may result in high-temperature strain and corrosion of the superheater tubes; also, the pressure below 40 bar lowers the requirements for pretreatment of the feed water [7,43,44].

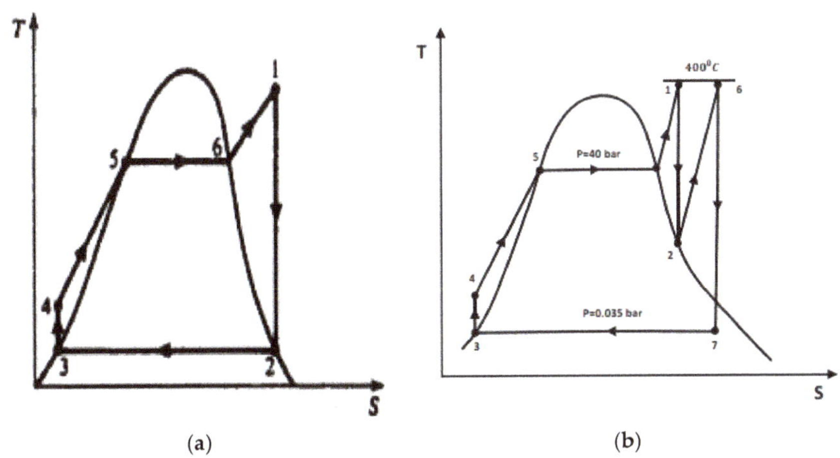

Figure 2. (**a**) T-S Rankine cycle with superheat; (**b**) T-S reheat Rankine cycle.

The reheat Rankine cycle in Figure 2b is a modification of simple Rankine cycle in Figure 2a. The schematic representation of a steam power plant using the reheat Rankine cycle is shown in Figure 3.

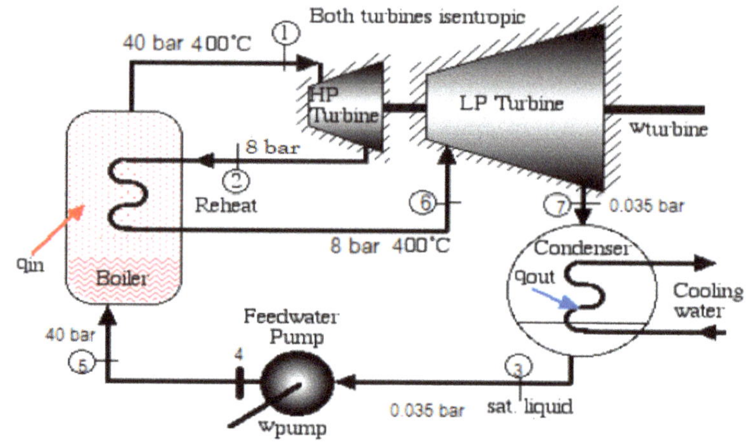

Figure 3. Rankine component layout of a steam power plant. Source: Karlsson and Jonsson [44].

Work Output ($W_{T_{12}}$) of the High-Pressure Turbine (HPT) can Be Determined Using Equation (9), as Suggested by Kaspooria [42], Akhator [12], Arabkooshar [45], and Loni [46]

$$\dot{W}_{T_{12}} = \dot{m}_{st}(h_1 - h_2) \tag{9}$$

where, $W_{T_{12}}$ (kJ/kg) is the HPT work output, $\dot{m}_{st}$ is the mass flow rate of the steam, h_1 and h_2 are the enthalpies at states 1 and 2 during the isentropic expansion process in the HPT.

According to Jordi [47] and Hesham [48], the Heat (Q_{26}) Supplied to the Steam at the Reheat Tube can be Determined Using Equation (10)

$$\dot{Q}_{T_{26}} = \dot{m}_{st}(h_6 - h_2) \tag{10}$$

where $\dot{Q}_{T_{26}}$ is the rate of heat supplied to the steam, during the constant pressure heating process, from state 2 to 6 in the reheat tube.

Work Output ($W_{T_{67}}$) of the Low-Pressure Turbine (LPT) can Be Determined as Suggested in Equation (11), by Akhator [12] and Arabkooshar [45]

$$\dot{W}_{T_{67}} = \dot{m}_{st}(h_6 - h_7) \tag{11}$$

where $W_{T_{67}}$ (kJ/kg) is the LPT work output, and h_6 and h_7 are the enthalpies at states 6 and 7 of the reheat Rankine cycle, during the isentropic expansion process in the LPT.

Heat ($\dot{Q}_{73}$) Rejected at the Condenser According to Akhator [12], Arabkooshar [45], and Loni [46] is Presented in Equation (12)

$$\dot{Q}_{73} = \dot{m}_{st}(h_7 - h_3) \tag{12}$$

where $\dot{Q}_{37}$ (kJ/kg) is the heat rejected in the condenser during an isothermal heat rejection process, while h_7 and h_3 are the enthalpies at states 7 and 3 of the reheat Rankine cycle respectively.

Heat ($\dot{Q}_{31}$) Supplied to the Boiler can Be Determined in Equation (13), as Suggested by Eastop [49], Sadhu [50], Ting [51], and Ibikunle [7]

$$\dot{Q}_{31} = \dot{m}_{st}(h_1 - h_3) \tag{13}$$

where $\dot{Q}_{31}$ (kJ/Kg) is the heat supplied to the boiler, while h_1 and h_3 are the enthalpies at states 1 and 3 of the reheat Rankine cycle respectively.

The Total Heat (Q_B) Supplied by the Boiler to the Steam

$$Q_B = Q_{31} + Q_{26} \tag{14}$$

where Q_B is the addition of heat Q_{13} and the heat of reheat Q_{26}.

2.6.1. The Cycle Efficiency (η_c)

According to Eastop [49], Sadhu [50], and Winterbone [52], the cycle efficiency is calculated using Equation (15):

$$\text{Cycle efficiency } \eta_c = \frac{\text{Work output}}{\text{Heat input}} = \frac{(W_{T_{12}} + W_{T_{67}})}{Q_B} \tag{15}$$

2.6.2. The Net Heat Supplied (Q_{net})

It is calculated by adopting Equation (16), as suggested by Eastop [49] and Sadhu [50]:

$$\text{Net heat supplied } (Q_{net}) = (Q_B - Q_{37}) \tag{16}$$

Estimation of the Capacity for the Components of the Steam Power Plant in Equations (17)–(22) Is Suggested by Ibrahim [53], Ibikunle [7], and Oyedepo [54]

$$\text{Steam mass flow rate } (\dot{m}_{st}) = \frac{Capacity\ of\ Steam\ power\ plant}{Net\ heat\ supplied} = \frac{EPP_{msw}}{Q_{net}} \tag{17}$$

$$\text{Specific steam consumption (SSC)} = \frac{3600}{W_{net}} \ (1\ \text{kW} = 3600\ \text{kJ/h}) \tag{18}$$

where net work (W_{net}) is equal to net heat (Q_{net}).

$$\text{MSW consumption rate, } \dot{m}_f = \dot{m}_{st}[((h_1 - h_3) + (h_6 - h_2))/(\eta B \times LHV)] \tag{19}$$

where m_f is taken as the rate of fuel consumption, the boiler efficiency is η_B, and the low heating value of MSW is LHV. Efficiency of 80% is assumed for the boiler (the range of boiler efficiency is 80–90%) [38].

$$\text{Boiler power } (Q_{BP}) : \text{ The boiler power, } Q_{BP} = \dot{m}_{st} \times Q_B \tag{20}$$

where Q_{BP} is considered as the power generated/day by the boiler for steam power plant.

$$\text{The turbine power output } (P_T) = \dot{m}_{st} \times W_T \tag{21}$$

The total work output of the turbine,

$$W_T = (W_{T_{12}} + W_{T_{67}}) \tag{22}$$

$$\text{The Condenser Power } (Q_{PC}) = \dot{m}_{st} \times Q_{73} \tag{23}$$

2.7. Development of Software for Selecting Power Plant Capacity

The nomenclature of all the parameters used in modeling the energy and electrical potentials of MSW, and the parameters used in modeling the power plant capacity required were prepared. An algorithm based on Equations (2)–(19) was prepared, which was translated into computer soft code using Java programming language (which is one of the common languages used in numerical analysis and a unified modeling language) [25]. The flow chart upon which the algorithm is based is shown in Figure 4. The software interface was developed using Java swing utilities which comprise mostly of JFrames, JTextrafields, JButtons, and JEditorpane for data display. Action listeners and action events were used to control the button functionalities, handle calculations, and panel/Editorpane displays. Java was used due to its flexibility, object-oriented programming language, and its multi-platform capabilities, which allows it to function on any operating system (OS). The conversion ratios for MSW in tons to kilograms, the heating value in MJ/kg to energy potential in kWh, the saturated and superheated water tables, and steam properties were incorporated into the software's database [25].

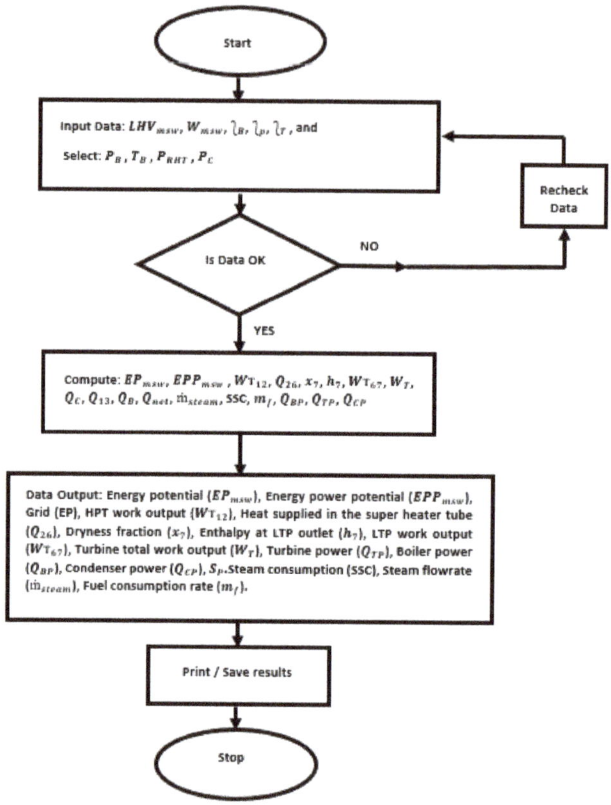

Figure 4. Flowchart of decision support system for steam power plant capacity.

The Flowchart for the Decision Support System

1. The weight of MSW available for energy generation, heating value of the MSW, boiler efficiency, conversion efficiency, and turbine efficiency was supplied into the corresponding designated box in the input interface shown in Figure 5,
2. The boiler pressure, boiler temperature, pressure in the reheat tube, and the pressure required in the condenser are selected from the corresponding draw-down of the icons provided in the input interface,
3. After the required parameters have been supplied/selected; the system requests that you proceed, and the proceed button is clicked,
4. Then, a verification interface, shown in Figure 6, will appear to ask if data is OKAY; if OK, click on proceed button otherwise click on the recheck data button and continue,
5. Then click on the process button, and it will automatically process the data. After that, click on proceed button and the result will automatically be displayed on an output interface, as well as the date and exact time of processing as presented in Figure 7.

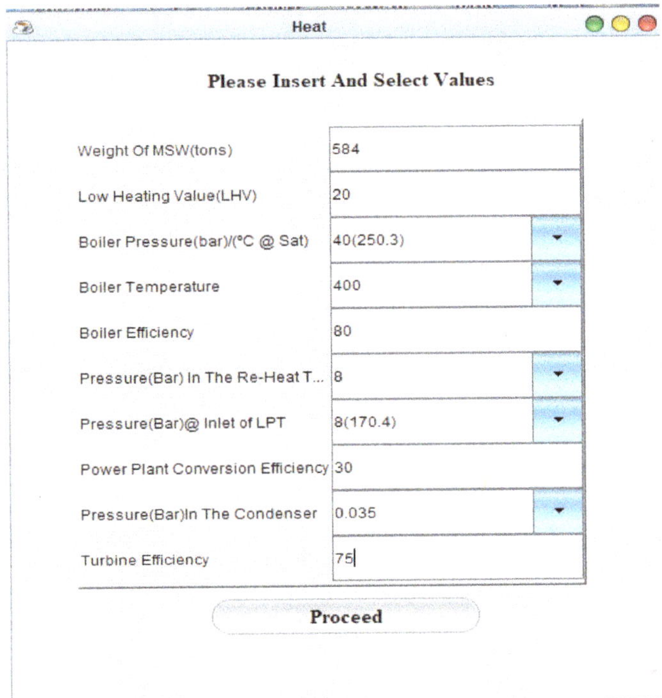

Figure 5. The input interface of the software to predict the capacity of the steam power plant.

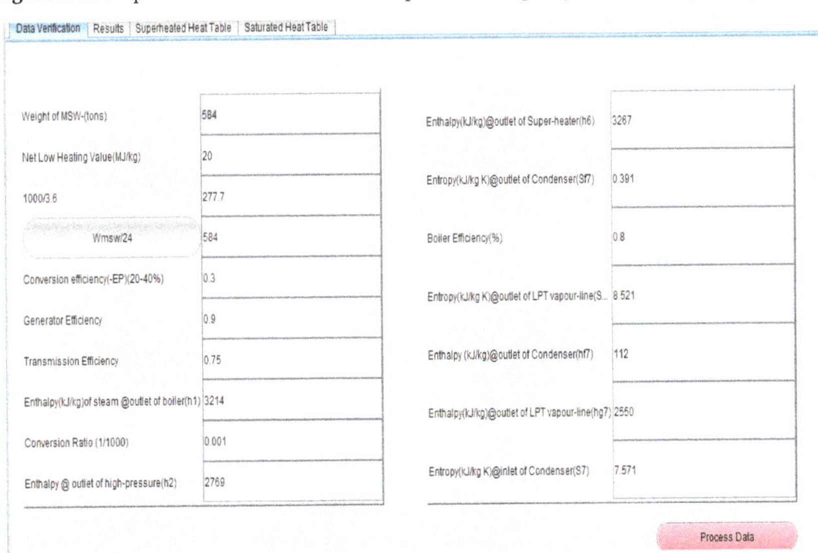

Figure 6. The verification interface of the software to recheck for the missing data.

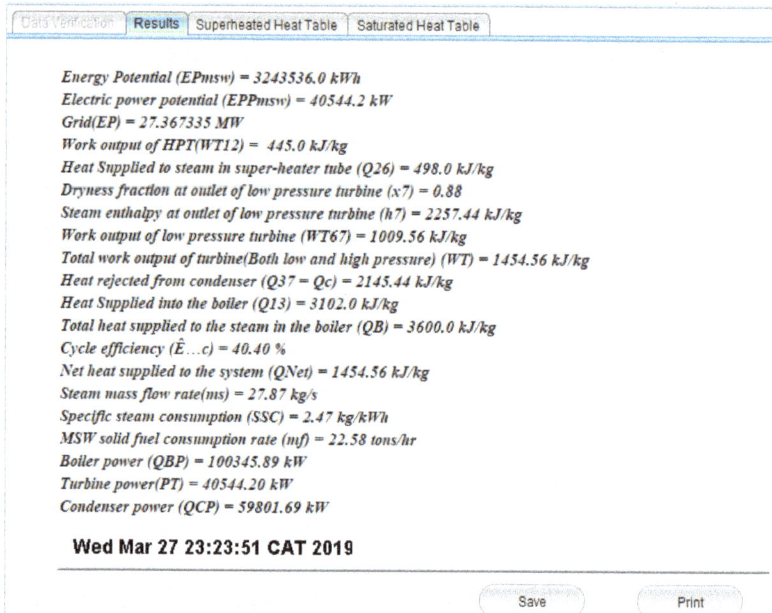

Figure 7. The values generated from the data processed by the software.

The results displayed include appropriate energy and electrical power potentials of the MSW, the work-output of the high-pressure and the low-pressure turbines, the heat rejected at the condenser, the heat supplied in the boiler, the boiler power, the turbine power, the condenser power required to convert the quantity of MSW to power, and the enthalpies at the different nodes of the reheat Rankine cycle.

3. Results and Discussion

In this section, the aggregate of the MSW produced in the Ilorin metropolis, the components in the MSW streams, their energy content, and the capacity of the waste-fired power plant required for electricity generation is established. Moreover, the manually-calculated values and the values generated by the selection support system are compared. These include the heat energy, electrical power potentials of a quantity of MSW, energy at the thermodynamic states of the Rankine cycle, and the required capacity of the power plant that will convert the MSW; this is to ascertain the accuracy of the decision support system.

3.1. The Estimated Aggregate of MSW Produced

The quantity of MSW generated in a city must be considered to be able to estimate the quantity of the fractions that are suitable for energy production. The MSW generated in Ilorin between January and June 2021, as presented in Table 1, was predicted to be about 101,912 tons, based on the facts and the figures of the waste collection system without the use of a weighing machine, compared to 370,706 tons/year generated in Onitsha (the eastern part of Nigeria) and 10,000 tons of MSW/day in Lagos, the commercial center of Nigeria [55], 250 tons of MSW/year in Ado-Ekiti (Western Nigeria) [56], 30 million tons/year in the US [57], 2.15 million tons/year in Sweden [58], 46 million in China [59], and 169,120 tons of MSW/day in Africa [60]. The aggregate waste produced during this study implies that the MSW generated in Ilorin is sufficient for the renewable energy process. The quantity collected for disposal was estimated, as suggested by Ogunjuyigbe [13], that the aggregate of MSW collected and disposed of in most of the developing nations is about 74% of the quantity generated.

Table 1. Estimation of MSW generated in Ilorin during the study based on transportation facts.

Collection Trucks	Trucks Used Per Day	Capacity (Tons)	Capacity (m^3)	Trips of Truck/Day	MSW Collected (Tons)	MSW Generated (Tons)
Dinno Tipper Truck	5	20	16	3	26,006.25	35,143.5
Hippo Tipper Truck	3	25	22	3	19,498.5	26,349
Scannia Compactor	2	30	22	2	10,405.5	14,061
Arm Roller	3	15	8	5	19,504.5	26,358
Total	13	90	68	13	75,414.75	101,911.5

3.2. The Physical Characterization of the MSW Components

Table 2 reveals that the quantity of MSW characterized during the study is about 985 kg. The population responsible for MSW generation during this study is 1,222,294 people as predicted by Ibikunle [5]. The aggregate waste characterized during June is about 229 kg, followed by 182 kg in May, and the least is about 156 kg in February. The reason for June having the highest quantity of MSW of about 23% of the total waste characterized, might be because during this period, newly harvested crops are emerging from farm and food items, and others are cheaper, making it affordable for many people than before. February had the least waste, which might be due to the dry season when there was no plant nor harvesting of food crops and people have just concluded the new year celebration and did not have enough money to purchase items. In this characterization study, plastic bottles have the highest fraction of about 13%, followed by grass/garden trimmings of about 8%, and leather having the least of about 1%. The reason for the high fraction of plastic bottles is because many people consume bottled water and other drinks. Leather has the lowest fraction because it is treasured for ornamental crafts, bags, and shoes in Ilorin, hence it is rare to come by its waste fractions. The generation rate of MSW during this study is 0.145 kg/capita/day.

Table 2. Physical characterization of MSW components (of 985 kg) for six months (January to February).

MSW Components	Jan.	Feb.	March	April	May	June	Total	Wt.	Kg/Capita
	Wt. (kg)	Wt. (kg)	Wt. (kg)	Wt. (kg)	Wt. (kg)	Wt. (kg)	Wt. (kg)	%	/Day
Food residue	12.25	10.00	8.35	8.00	14.21	26.35	79.16	8.04	0.012
Wood	4.65	4.85	4.25	3.85	3.25	2.45	23.30	2.37	0.003
Paper	8.80	6.50	6.20	4.60	8.35	6.55	41.00	4.16	0.006
Packaging box	14.24	18.55	12.50	7.80	10.55	12.20	75.84	7.70	0.011
Grass/trimmings	3.64	8.20	10.60	12.40	21.12	24.22	80.18	8.14	0.012
Textiles (rags)	4.20	6.22	8.46	10.20	10.25	14.24	53.57	5.44	0.008
Toiletries	3.40	4.56	2.28	1.12	0.84	3.18	15.38	1.56	0.002
Feces	1.64	4.62	2.08	5.10	8.24	10.32	32.00	3.25	0.005
Cow dung	8.45	6.45	5.45	3.35	4.42	4.45	32.57	3.31	0.005
Nylon (water sachet)	18.00	16.24	10.24	8.46	6.45	8.22	67.61	6.87	0.010
Polypropylene sack	12.20	8.42	6.20	8.20	10.55	20.14	65.71	6.67	0.010
Plastic bottle	30.25	22.00	20.68	14.24	18.24	20.12	125.53	12.75	0.018
Rubber	0.82	1.20	1.22	0.14	5.20	5.24	13.82	1.40	0.002
Leather	0.64	0.26	1.20	0.22	4.10	4.16	10.58	1.07	0.002
Glass/Ceramics	10.40	9.80	10.40	6.00	10.12	9.85	56.57	5.74	0.008
Animal bones	2.64	1.12	0.80	2.45	10.22	12.25	29.48	2.99	0.004
Tins/Metals	12.00	10.00	14.00	8.20	6.35	14.66	65.21	6.62	0.010
Sand/Ash	10.90	8.80	6.30	4.30	5.12	4.62	40.04	4.07	0.006
Other biogenic	6.24	8.64	5.40	6.80	24.54	25.52	77.14	7.83	0.011
Grand total	165.36	156.43	136.61	115.43	182.12	228.74	984.69	100.00	0.145

3.3. The Combustible MSW Fractions That Are Considered for Energy Production

Nine combustible fractions of the 19 MSW components that were characterized are considered for energy production, as presented in Table 3. The table shows that about 612 kg of combustible waste fractions were characterized during this study. This makes about 62% of the waste generated. This implies about 63,185 tons of combustible MSW were available for energy production during the investigation. Ibikunle et al. (2019) estimated the combustible components of MSW in Ilorin for the year 2016 to be about 71%, compared to 84% in Ghana in 1983 and 27% in 2014; 7.7% in South Africa; 74% in Nigeria, and 3.6% of the US [61]. This implies energy recovery from MSW via combustion in 2016 will reduce the quantity of waste deposited into the dumpsite in Ilorin metropolis by about 71%, but this recent study revealed that energy recovery via combustion of waste fractions could reduce the aggregate waste deposited into the dumpsite in Ilorin by about 62% within six months.

Table 3. Physical characterization of MSW (of 612 kg) for energy production between January–February.

MSW Components	Jan.	Feb.	March	April	May	June	Total	Wt.
	Wt. (kg)	Wt. (kg)	Wt. (kg)	Wt. (kg)	Wt. (kg)	Wt. (kg)	Wt. (kg)	%
Food residue	12.25	10.00	8.35	8.00	14.21	26.35	79.16	8.04
Wood	4.65	4.85	4.25	3.85	3.25	2.45	23.30	2.37
Paper	8.80	6.50	6.20	4.60	8.35	6.55	41.00	4.16
Packaging box	14.24	18.55	12.50	7.80	10.55	12.20	75.84	7.70
Grass/garden trimmings	3.64	8.20	10.60	12.40	21.12	24.22	80.18	8.14
Textiles (rags)	4.20	6.22	8.46	10.20	10.25	14.24	53.57	5.44
Nylon (water sachet)	18.00	16.24	10.24	8.46	6.45	8.22	67.61	6.87
Polypropylene sack	12.20	8.42	6.20	8.20	10.55	20.14	65.71	6.67
Plastic bottle	30.25	22.00	20.68	14.24	18.24	20.12	125.53	12.75
Total	108.23	100.98	87.48	77.75	102.97	134.49	611.90	62.14

3.4. The Thermochemical Analysis of Combustible MSW Components

Table 4 presents the ultimate analysis of the waste fractions and the high heating value (HHV) of the components obtained from the bomb calorimeter. The analyses reveal that the average elemental components of the waste fractions give 29.14% composition of carbon, 0.11% of hydrogen, 3.95% of nitrogen, 0.50% of sulfur, and 0.15% of oxygen. The average HHV of the MSW components is 25 MJ/kg. These values are very important in the estimation of the energy and power potentials of MSW.

Table 4. Thermochemical analysis of MSW components.

MSW Fractions	C%	H%	N%	S%	O%	HHV (MJ/kg)
Food residue	37.88 ± 0.44	0.17 ± 0.00	4.68 ± 0.05	3.80 ± 0.02	0.16 ± 0.00	18.56
Wood	36.80 ± 0.62	0.11 ± 0.02	4.46 ± 0.04	0.06 ± 0.02	0.08 ± 0.00	18.38
Paper	34.68 ± 0.08	0.08 ± 0.02	4.58 ± 0.24	0.10 ± 0.00	0.07 ± 0.00	17.28
Packaging box	21.45 ± 0.06	0.11 ± 0.00	3.32 ± 0.10	0.08 ± 0.03	0.12 ± 0.00	16.68
Grass/garden trimmings	31.46 ± 0.11	0.09 ± 0.02	3.81 ± 0.06	0.07 ± 0.00	0.06 ± 0.00	17.78
Textiles (rags)	33.66 ± 0.14	0.10 ± 0.00	4.26 ± 0.00	0.14 ± 0.00	0.15 ± 0.00	16.45
Nylon	21.86 ± 0.08	0.11 ± 0.00	3.56 ± 0.02	0.13 ± 0.00	0.08 ± 0.00	45.26
Polypropylene sack	21.65 ± 0.11	0.11 ± 0.02	3.42 ± 0.01	0.06 ± 0.00	0.06 ± 0.00	38.88
Plastic bottle	22.84 ± 0.03	0.10 ± 0.00	3.43 ± 0.00	0.08 ± 0.02	0.11 ± 0.02	36.86

The low heating value (LHV) of the waste components that is paramount in the estimation of power potential as determined by adopting Dulong and Steuer models in Equations (3) and (4) is presented in Table 5a. The average HHV from the bomb calorimeter is 25 MJ/kg, the average LHV^b from the Dulong model is about 21 MJ/kg, and LHV^c from the Steuer model is about 20.8 MJ/kg. Moreover, the LHV obtained from Equation (2) is about 19 MJ/kg. Therefore, considering the results from the three equations, the typical value for the LHV is assumed to be 20 MJ/kg. The MSW with LHV of 20 MJ/kg will produce correspondent energy of about 43% of petrol, 45% of energy in diesel, 48% of that in natural gas, 49% in coal, and about 100% of energy content in woody biomass [62].

Table 5. (a) The LHV of the MSW components from Dulong and Steuer models; (b) Energy potential (EP_{MSW}) and electrical power potential (EPP_{MSW}) of MSW.

(a)

MSW Fractions	HHV (MJ/kg)	HV%	LHV[b] (MJ/kg)	LHV[c] (MJ/kg)	Typical LHV (MJ/kg)
Food residue	18.56	8.21	27.20	26.50	26.85
Wood	18.38	8.13	24.12	24.31	24.22
Paper	17.28	7.64	26.03	26.00	26.02
Packaging box	16.68	7.38	16.01	16.01	16.01
Grass/garden trimmings	17.78	7.86	21.03	21.51	21.27
Textiles (rags)	16.45	7.27	24.21	23.34	23.78
Nylon	45.26	20.02	17.02	18.10	17.56
Polypropylene sack	38.88	17.19	15.25	17.02	16.14
Plastic bottle	36.86	16.30	15.16	17.00	16.08
Total	226.13	100	186.03	189.79	187.91
Average	25		20.67	21.09	20.88

(b)

Types	MSW (tons)/day	LHV (MJ/kg)	EP_{MSW} (kWh)	EPP_{MSW} (kW)
Food residue	85.75	1.93	263,067	3288.34
Wood	6.62	0.15	265,084	3313.55
Paper	49.95	1.03	244,915	3061.44
Packaging box	80.14	1.54	229,067	2863.34
Grass/garden trimmings	36.97	0.80	256,440	3205.50
Textiles (rags)	73.69	1.40	226,186	2827.33
Nylon	125.95	7.03	665,592	8319.90
Polypropylene sack	43.99	2.09	561,863	7023.29
Plastic bottle	80.96	3.65	535,931	6699.14
Total	584	20	3,248,145	40,601.83

HHV: the high heating value from calorimetry experiment; LHV[b]: the low heating value from Dulong model; LHV[c]: the low heating value from Steuer model, and Typical LHV: the average of the values from the models.

3.5. The Energy Potential (EP_{MSW}) and Electrical Power Potential of MSW (EPP_{MSW})

The methodology established that about 62% of the MSW characterized are combustible, therefore, considering about 584 tons out of about 63,185 tons of (combustible) waste fractions available during the study, with the low heating value (LHV) of 20 MJ/kg, it can produce heat energy potential of 3.2 GWh and electrical power potential of 40.6 MW in a day, as presented in Table 5b. The available energy potential (EP_{MSW}) of 3.2 GWh in MSW will produce equivalent energy in 603 tons of wood, 395 tons of coal, 94 tons of hydrogen, 352,591 L of petrol, and 320,000 L of diesel [63]. The electrical power potential (EPP_{MSW}) of the MSW is about 41 MW, and this will meet about 15% of the power requirement in Kwara State [64], as well as will help achieve about 9% of the Nigeria Renewable Energy Master Plan (REMP) goal for 2025.

3.6. The Capacity of the Steam Power Plant Components Required to Convert the MSW

The capacities of the steam power plant components that will convert the available combustible MSW in Ilorin to energy were determined. It is established that a steam power plant output of 41 MW, operating on a reheat Rankine cycle of 40% efficiency, based on design specifications which operate in steam pressure and temperature of 40 bar and 400 °C respectively is required. Temperature higher than 400 °C can bring about high-temperature strain and corrosion in the boiler superheater tubes, and pressure below 40 bar can lower the requirements for pretreatment of the feed water [27,44,45]. The plant has an MSW fuel consumption rate of 40 tons/h, boiler capacity of 3600 kJ/kg (heat input), turbine output capacity of 1455 kJ/kg of steam, steam mass flow rate of 28 kg/s (100 tons/h), and specific steam consumption (SSC) of 2.48 kg/kWh.

3.7. The Thermodynamic Properties at each State in the Power Plant Cycle and the Capacities of the Power Plant Components

The parameters determined for each node (state) during the thermodynamic processes required in the steam power generating procedures are given in Table 6a. The energy balance of the processes is given in Table 6b and the capacities for each equipment of the power plant required are given in Table 6c. Table 6a,b contain pressures and temperatures selected from the steam table incorporated into the power plant design calculations to determine the enthalpy, entropy, work done, and the heat produced in different nodes of thermodynamic stages.

Table 6. (a) The parameters determined for the nodes of the steam power plant cycle; (b) The work done and the heat produced during the Rankine processes; (c) The capacities designed for the steam boiler power plant components.

(a)

Description	Node	Pressure (bar)	Temperature (°C)	Enthalpy (kJ/kg)	Entropy (kJ/kg K)
Boiler outlet/H-P turbine inlet	1	40	400	3214	6.769
H-P turbine outlet/intermediate superheater inlet	2	8	170.42	2769	6.663
Intermediate superheater outlet/L-P turbine inlet	6	8	400	3267	7.571
L-P turbine outlet/condenser inlet	7	0.035	26.7	2257.44	7.571
Condenser outlet/pump inlet	3	0.035	26.7	112	0.391
Pump outlet/boiler inlet	4	40	205	875	2.382

(b)

Work done	kJ/kg	Heat	kJ/kg
H-P turbine (WT_{12})	445	Heat supplied into boiler (Q_{13})	498
L-P turbine (WT_{67})	1010	Heat supplied during reheat (Q_{26})	498
Total work output of turbine (WT)	1455	Heat rejected from condenser (Q_{37})	2145
Net work done (W_{net})	1455	Net heat supplied (Q_{net})	1455

(c)

Equipment	Heat (kJ/kg)	Work (kJ/kg)	Power (MW)
Boiler	3600	–	101
Turbine	–	1455	41
Condenser	2145	–	60
Steam and Fuel Consumption Rate of The Plant			
Water/fuel consumption	kg/s	(tons/h)	(kg/kWh)
Steam mass flow rate ($\dot{m}_s$)	27.87	100	
Fuel consumption rate (m_f)	6.25	22.58	
Specific steam consumption (SSC)			2.48

Table 6c presents the heat, work, and power estimated for boiler, turbine, and condenser discretely, while determining the capacity of the power plant that will convert 584 tons of MSW to energy. This agrees with 'ZGB' (a manufacturer of boilers in China) with a rating range of 101–150 tons/h steam mass flow rate for boiler with the power of 70–105 MW.

3.8. Software-Generated Values on Power Potential and Power Plant Capacity

The software developed serves as the supporting tool for the selection of waste-to-energy (WTE) power plant capacity that will utilize MSW as fuel. This system determines the energy and power potentials of MSW when the known weight of MSW in tons and its low heating value (MJ/kg) are inputted into the input interface of the software shown in Figure 5. The data inputted are verified in the interface in Figure 6, to ensure error-free processes, before the data is processed. The results obtained using the software are presented in output display Figure 7. This decision support system will predict the capacity of the power plant required for energy recovery of any quantity of MSW at any location, as well as the heat enthalpy at the nodes of the Rankine cycle during the thermodynamic processes, provided the quantity of waste to combust is known, as well as the value of the low heating value.

3.9. Comparison between Manually-Calculated and Software-Generated Values on: Energy Potentials and Power Plant Capacity

The appropriate way to validate the efficiency of the software developed can be validated by comparing the manually-calculated values to the values obtained from the software package [65]. The values presented in Table 7 show that the accuracy of the software is >99%. This software can be used to predict the energy and power potentials of MSW in any location at any time, provided the weight and the low heating value are known. Nevertheless, it will excellently predict the quantity of work or heat involved in each of the power plant's equipment, as well as the capacity of the required boiler, turbine, and condenser that will convert the MSW to electrical energy.

Table 7. Comparison of manually-calculated and software-generated values.

Parameters	Manually-Calculated Values	Software-Generated Values
Energy potential (EP_{msw})	3,243,536.0 kW	3,243,536.0 kW
Electrical power potential (EPP_{msw})	40,544.3 kW	40,544.3 kW
Power to grid (GP)	27.36732 MW	27.36733 MW
Work output of the high-pressure turbine ($W_{T_{12}}$)	445.0 kJ/kg	445.0 KJ/kg
Heat supplied during reheat process (Q_{26})	498.0 kJ/kg	498.0 kJ/kg
Dryness fraction at outlet of low-pressure turbine (x_7)	0.88	0.88
Steam enthalpy at outlet of low-pressure turbine (h_7)	2257.44 kJ/kg	2257.44 kJ/kg
Work output of low-pressure turbine ($W_{T_{67}}$)	1009.56 kJ/kg	1009.56 kJ/kg
Total work output of turbine ($W_{T_{12}} + W_{T_{67}}$), W_T	1454.56 kJ/kg	1454.56 kJ/kg
Heat rejected from condenser (Q_{37}) = Q_C	2145.44 kJ/kg	2145.44 kJ/kg
Heat supplied into the boiler (Q_{13})	3102.0 kJ/kg	3102.0 kJ/kg
Total heat supplied to the steam in the boiler (Q_B)	3600.0 kJ/kg	3600.0 kJ/kg
Cycle efficiency (S_c)	40.40%	40.40%
Net heat supplied to the system (Q_{Net})	1454.56 kJ/kg	1454.56 kJ/kg
Steam mass flow rate ($\dot{m}_s$)	27.87 kg/s	27.87 kg/s
Specific steam consumption (SSC)	2.48 kg/kWh	2.47 kg/kWh
MSW (fuel) consumption rate (m_f)	22.58 tons/h	22.58 tons/h
Boiler power (Q_{BP})	100,345.88 kW	100,345.89 kW
Turbine power (Q_{TP})	40,540.20 kW	40,544.20 kW
Condenser power (Q_{CP})	59,801.68 kW	59,801.68 kW

4. Conclusions

The software developed can effectively serve as a supporting system for the selection of the appropriate capacity of the steam power plant to convert any quantity of MSW as fuel to energy. The results obtained revealed that a steam power plant of about 100 MW power of boiler, with about 40 MW power of turbine and 60 MW power of condenser with a steam mass flow rate of about 28 kg/s, will convert about 584 tons of municipal solid waste (MSW) with 20 MJ/kg heating value to electrical power potential of about 41 MW, and energy potential of about 3243 MWh. The comparison of the results obtained by manual calculation and that of the software shows an accuracy greater than 99%. It is concluded that the software will successfully predict the energy and electrical power potentials of MSW, and the power to grid potentials in any quantity of combustible waste fractions at any time, irrespective of geographical variations. It will also predict the capacities of individual components in the power plant required to convert the waste to energy, provided the quantity of the municipal solid waste (MSW) to be converted in tons and the net heating value of the MSW (MJ/kg) is supplied. The software will also determine the cycle efficiency of the power plant, the steam mass flow rate, and the MSW (fuel) consumption.

Author Contributions: R.A.I. collected the literature and data, developed the software, inputted data into the software and prepared the manuscript; I.F.T. interpreted the data, analyzed the results, and designed the power plant capacity; B.O.A. prepared the algorithm used to develop the soft code of the software, supervised the experimentation, and edited the manuscript text. All authors have read and agreed to the published version of the manuscript.

Funding: This research was funded by the authors without any financial aid from an external body.

Acknowledgments: The authors appreciate the technologists in the research laboratories of Landmark University and the Federal University of Technology Akure for their support during experimentation.

Conflicts of Interest: The authors declare no conflict of interest of any kind.

References

1. Omari, A.; Said, M.; Njau, K.; John, G.; Mtul, P. Energy Recovery routes from Municipal Solid Waste, A case study of Arusha-Tanzania. *J. Energy Technol. Policy* **2014**, *4*. Available online: https://www.iiste.org/Journals/index.php/JETP/article/view/13157 (accessed on 7 May 2021).
2. Titiladunayo, I.F.; Akinnuli, B.O.; Ibikunle, R.A.; Agboola, O.O.; Ogunsemi, B.T. Analysis of combustible municipal solid waste fractions as fuel for energy production: Exploring its physicochemical and thermal characteristics. *Int. J. Civil. Eng. Technol.* **2018**, *9*, 1557–1575. Available online: http://www.iaeme.com/MasterAdmin/Journal_uploads/IJCIET/VOLUME_9_ISSUE_13/IJCIET_09_13_157.pdf (accessed on 20 December 2018).
3. Islam, K.M.N. Municipal solid waste to energy generation in Bangladesh: Possible scenarios to generate renewable electricity in Dhaka and Chittagong city. *J. Renew. Energy* **2016**, 1–16. Available online: http://www.hindawi.com/journals (accessed on 25 November 2017).
4. Ibikunle, R.A.; Titiladunayo, I.F.; Lukman, A.F.; Dahunsi, S.O.; Akeju, E.A. Municipal solid waste sampling, quantification, and seasonal characterization for power evaluation: Energy potential and statistical modelling. *Fuel* **2020**, *277*, 118122. [CrossRef]
5. Ibikunle, R.A. Investigating municipal solid waste generation and management in Ilorin for possible integrated waste-management system. *J. Mater. Cycles Waste Manag.* **2021**, *23*, 1239–1257. [CrossRef]
6. Johari, A.; Hashim, H.; Mat, R.; Alias, H.; Hasshim, M.H.; Rozainee, M. Generalization, formulation, and heat contents of simulated MSW with moisture content. *J. Eng. Sci. Technol.* **2012**, *7*, 701–710. Available online: http://jestec.taylors.edu.my/Vol%207%20Issue%206%20December%2012/Vol7_.2012 (accessed on 5 May 2021).
7. Ibikunle, R.A.; Titiladunayo, I.F.; Akinnuli, B.O.; Dahunsi, S.O.; Olayanju, T.M.A. Estimation of power generation from municipal solid wastes: A case study of Ilorin metropolis. *Nigeria. Energy Rep.* **2019**, *9*, 126–135. [CrossRef]
8. Hoornweg, D.; Bhada-Tata, P. *What a Waste: A Global Review of Solid Waste Management*; Urban Development Series; Knowledge Papers no. 15; World Bank: Washington, DC, USA, 2012. Available online: https://openknowledge.worldbank.org/handle/10986/17388 (accessed on 5 May 2021).
9. Parashar, C.K.; Das, P.; Samanta, S.; Ganguly, A.; Chatterjee, P.K. Municipal Solid Wastes—A Promising Sustainable Source of Energy: A Review on Different Waste-to-Energy Conversion Technologies. Energy Recovery Processes from Wastes. 2019, pp. 151–163. Available online: https://link.springer.com/chapter/10.1007/978-981-32-9228-4_13 (accessed on 5 May 2021).
10. Ibikunle, R.A.; Titiladunayo, I.F.; Dahunsi, S.O.; Akeju, E.A.; Osueke, C.O. Characterization and projection of dry season municipal solid waste for energy production in Ilorin metropolis, Nigeria. *J. Waste Manag. Res.* **2021**, 1–10. [CrossRef]
11. World Bank. Sustainable Energy for All Database from SE4ALL Global Tracking Framework Led Jointly by the World Bank International Energy Agency, and Energy Management Assistance Programme. Available online: https://data.worldbank.org (accessed on 24 April 2018).
12. Akhator, E.P.; Obonor, A.I.; Ezemonye, L.I. Electricity generation in Nigeria from municipal solid waste using the Swedish Waste-to-Energy Model. *J. Appl. Sci. Environ. Manag.* **2016**, *20*, 635–643. [CrossRef]
13. Ogunjuyigbe, A.S.O.; Ayodele, T.R.; Alao, M.A. Electricity generation from municipal solid waste in some selected cities of Nigeria: An assessment of feasibility, potential technologies. *J. Renew. Sustain. Energy Rev.* **2017**, *80*, 149–162. [CrossRef]
14. Rasmus, L.; Brian, V.M. Large combined heat and power plants in sustainable energy systems. *Appl. Energy* **2015**, *142*, 389–395. Available online: https://ideas.repec.org/a/eee/appene/v142y2015icp389-395.html (accessed on 4 April 2021).
15. Tsolas, S.D.; Karim, M.N.; Hasan, M.M.F. Optimization of water-energy nexus: A network representation-based graphical approach. *Appl. Energy* **2018**, *224*, 230–250. [CrossRef]
16. Xuerui, G.; Yong, Z.; Shibao, L.; Qianyun, C.; Tingli, A.; Xinxueqi, H.; La, Z. Impact of coal power production on sustainable water resources management in the coal-fired power energy bases of Northern China. *Appl. Energy* **2019**, *250*, 821–833.

17. Do, W.K.; Tong, S.K.; Kwang, B.H.; Jung, K.P. The effect of firing biogas on the performance and operating characteristics of simple and recuperative cycle gas turbine combined heat and power systems. *Appl. Energy* **2012**, *93*, 215–228.
18. Njoku, N.; Lamond, J.; Everett, G.; Manu, P. An overview of municipal solid waste management in developing and developed economies: Analysis of practices andcontributions to urban flooding in Sub-Saharan Africa. In Proceedings of the 12th International Post Graduate Research Conference, Media City, UK, 10–12 June 2015. [CrossRef]
19. Ibikunle, R.A.; Lukman, A.F.; Titiladunayo, I.F.; Akeju, E.A.; Dahunsi, S.O. Modeling and robust prediction of high heating values of municipal solid waste based on ultimate analysis. *Energy Sources Part A Recovery Util. Environ. Effects* **2020**. [CrossRef]
20. Worldometer-Real Time Statistics (Population by Country). Available online: https://www.worldometers.info2019 (accessed on 3 May 2019).
21. Hongtao, W.; Yongfeng, N. Municipal Solid Waste Characteristics and Management in China. *J. Air Waste Manag. Assoc.* **2001**, *51*, 250–263. [CrossRef]
22. Kosuke, K.; Tomohiro, T. Revisiting estimates of municipal solid waste generation per capita and their reliability. *J. Mater. Cycles Waste Manag.* **2016**, *18*, 1–3.
23. U.S. Energy Information Administration (U.S. EIA). Electricity in the United States is Produced (Generated) with Diverse Energy Sources and Technologies. 2020. Available online: https://www.eia.gov/energyexplained/electricity/electricity-in-the-us.php (accessed on 4 July 2020).
24. United Nations Environmental Programme (UNEP). Euthopia Marching towards Africa's First Waste-to-Energy Plant: Africa News. 2017. Available online: http://www.africanews.com/2017/11/25/ethiopia-marching-towards-africas-first-waste-to-energy-plant-unep (accessed on 20 December 2018).
25. Oladende, O. 12 Nigerian Power Plants Produced Zero Megawatt Electricity on Christmas Day. Premiums Times (News). 4 January 2017. Available online: https://www.premiumtimesng.com/news/headlines/219590-12-nigerian-power-plants-produced-zero-megawatt-electricity-christmas-day.html (accessed on 25 June 2021).
26. Larmuth, J.; Cuellar, A. An updated review of South African CSP projects under the renewable energy independent power producer procurement programme (REIPPPP). In *AIP Conference Proceedings*; AIP Publishing LLC.: Melville, NY, USA, 2019; Volume 2126, p. 040001. [CrossRef]
27. Ibikunle, R.A. Decision Support System for Power Plant Design Using Available Municipal Solid Waste in Ilorin Metropolis as Fuel. Ph.D. Thesis, Federal University of Technology, Akure, Nigeria, 2019.
28. Mareck, J.D.; Roger, R.F. Decision Support Systems. In *Encyclopedia of Library and Information Sciences*, 2nd ed.; Kent, A., Ed.; Marcel Dekker, Inc.: New York, NY, USA, 2002.
29. Claudio, B. Decision Support Systems Development: A Methodological Approach. *J. Appl. Bus. Inf. Syst.* **2011**, *2*, 151–158. Available online: http://www.jabis.ro/2011/4/0204012011.pdf (accessed on 1 May 2021).
30. Konstantinos, I.; Georgios, T.; Garyfallos, A.; Zacharoula, A.; Eleni, Z. A Spatial Decision Support System Framework for the Evaluation of Biomass Energy Production Locations: Case Study in the Regional Unit of Drama Greece. *Sustainability* **2018**, *10*, 531–553. [CrossRef]
31. Fatima, C.C.D.; Erhard, W.P. A decision support system for power plant design. *Eur. J. Oper. Res.* **1998**, *109*, 310–320. Available online: http://www.sciencedirect.com/science/article/pii/S0377-2217(98)00059-91998 (accessed on 5 July 2019).
32. Konstantinos, I.; Paragnotis, L.; Arabatzis, Z. Development of a decision support system for the study of an area after the occurrence of forest fire. *Int. J. Sustain. Soc.* **2011**, *3*, 5–53. [CrossRef]
33. NT ENVR-01 (Nordtest Method). Solid Waste Municipal Sampling and Characterization. Published by Nordtest Tekniikantie 12, FIN–02150 ESPOO FINLAND. 1998. Available online: http://www.nordtest.info/wp/1995/05/14/solid-waste-municipal-sampling-and-characterisation-nt-envir-001/ (accessed on 7 July 2019).
34. Issam, A.K.A.; Maria, M.B.; Salam, A.T.; Saheed, H.Q.; Kassinos, D.B. Solid waste characterization, qualification, and management practices in developing countries, a case study: Nubulus district-palestine. *J. Environ. Manag.* **2010**, *91*, 1131–1138.
35. Vairam, S.; Ramesh, S. *Engineering Chemistry*; John Wiley and Sons Ltd.: Southern Gate, Chichester, UK, 2013.
36. Shi, H.; Maphinpey, N.; Aqsha, A.; Silbermann, R. Characterization, thermochemical conversion studies and heating value modeling of municipal solid waste. *J. Waste Manag.* **2016**, *48*, 34–47. [CrossRef]
37. Ibikunle, R.A.; Titiladunayo, I.F.; Akinnuli, B.O.; Lukman, A.F.; Ikubanni, P.P.; Agboola, O.O. Modeling the energy content of municipal solid waste and determination of its physicochemical correlation using multiple regression analysis. *Int. J. Mech. Eng Technol.* **2018**, *9*, 220–232.
38. Kumar, J.S.; Subbaiah, K.V.; Rao, P.P. Waste to energy-a case study of eluru city, andhara pradesh. *Int. J. Environ. Sci.* **2010**, *2*, 151–162.
39. Daura, L.A. Electricity generation potential of municipal solid waste in kanometropolis. *J. Sci. Eng. Res.* **2016**, *3*, 157–161.
40. Muhammad, A.; Farid, N.A.; Ab Saman, K. The energy potential of municipal solid waste for power generation in indonesia. *J. Mech.* **2014**, *2*, 42–54.
41. Bright Hub Engineering (BHE). Conversion Efficiencies in the Steam Power Plants. 2016. Available online: www.brighthubengineering.com/power-plants/72369-compare-thefficiency (accessed on 22 February 2017).
42. Kapooria, R.K.; Kumar, S.; Kasana, K.S. An analysis of a thermal power plant working on a Rankine cycle: A theoretical investigation. *J. Energy S. Afr.* **2008**, *19*, 77–83. [CrossRef]

43. Somplak, R.; Ferdan, T.; Palvas, M.; Palvas, M.; Popela, P. Waste-to-energy facility planning under uncertain circumstances. *J. Appl. Therm. Eng.* **2013**, *61*, 106–114. [CrossRef]
44. Karlsson, L.; Jonsson, T.L. Pre-Feasibility Study of a Waste-to-Energy Plant in Chisinau, Moldova. Uppsala Universitet. 2012. Available online: https://www.diva-portal.org/smash/get/diva2:510286/FULLTEXT01.pdf (accessed on 25 October 2016).
45. Arabkoohsar, A. Combined steam based high-temperature heat and power storage with an Organic Rankine Cycle, an efficient mechanical electricity storage technology. *J. Clean. Prod.* **2020**, *247*, 119098. [CrossRef]
46. Loni, R.; Najafi, G.; Bellos, E.; Rajaee, F.; Said, F.; Mazlan, M. A review of industrial waste heat recovery system for power generation with Organic Rankine Cycle: Recent challenges and future outlook. *J. Clean. Prod.* **2020**, 125070. [CrossRef]
47. Jordi, B. Development and Implementation of a Nuclear Power Plant Steam Turbine Model in the System Code ATHLET. Master Thesis, Technische Universitat Muchen, Muchen, Germany, 2011.
48. Hesham, G.I. Steam Power Plant Design Upgrading (Case Study: Khoms Steam Power Plant). *J. Energy Environ. Res.* **2011**, *1*, 202. [CrossRef]
49. Eastop, T.D.; Mc Conkey, A. *Applied Thermodynamics*, 5th ed.; Dorling Kindersley Pvt. Ltd.: Delhi, India, 2014.
50. Sadhu, S. *Handbook of Mechanical Engineering*; S. Chand & Company Ltd.: New Delhi, India, 2011.
51. Ting, Y.; Noam, L. Thermodynamic analysis of hybrid Rankine cycles using multiple heat sources of different temperatures. *Appl. Energy* **2018**, *222*, 564–583. [CrossRef]
52. Winterbone, O.E.; Turan, A. *Advanced Thermodynamics for Engineers*; John Wiley and Sons Inc.: New York, NY, USA; Toronto, ON, Canada, 2015.
53. Ibrahim, D.; Yusuf, B. Fundamentals of Energy System. Integrated Energy Systems for Multigeneration (Elsevier Inc.). 2019. Available online: https://www.elsevier.com/books/integrated-energy-systems-formultigeneration/dincer/978-0-12-809943-8 (accessed on 7 May 2021). [CrossRef]
54. Oyedepo, S.O.; Fakeye, B.A.; Mabinuori, B.; Babalola, P.O.; Leramo, R.O.; Kilanko, O.; Dirisu, J.O.; Udo, M.; Efemwenkiekie, U.K.; Oyebanji, J.A. Thermodynamics analysis and performance optimization of a reheat–Regenerative steam turbine power plant with feed water heaters. *Fuel* **2020**, *280*, 118577. [CrossRef]
55. Maxwell, U.N. Solid waste generation and disposal in a nigerian city: An empirical analysis in onitsha metropolis. *J. Environ. Manag. Saf.* **2010**, *1*, 180–191.
56. Rominiyi, O.L.; Fapetu, O.P.; Owolabi, J.O.; Adaramola, B.A. Determination of energycontent of the municipal solid waste of Ado-Ekiti Metropolis, Southwest, Nigeria. *Curr. J. Appl. Sci. Technol.* **2017**, *23*, 1–11. [CrossRef]
57. U.S. Energy Information Administration (USEIA). Biomass Explained. Waste-to-Energy (Municipal Solid Waste) Energy Recovery from Municipal Solid Waste. 2018. Available online: https://www.eia.gov/state/seds/sep_fuel/notes/use_glossary.pdf (accessed on 7 July 2019).
58. Swedish Waste Management (SWM). 2015. Available online: www.arfallaverage.se/fileadmin/uploads/Rapporter/SWM2015.pdf (accessed on 20 June 2016).
59. Scarlet, W.; Dellemand, J.F.; Monforti-Ferrano, F.; Nita, V. The role of biomass and bioenergy in a future bioeconomy: Policies and facts. *Environ. Dev.* **2015**, *15*, 3–34. [CrossRef]
60. Wale, B. Energy. Energy Consult Powering Clean Energy Future, Solid Waste Management in Nigeria. 2018. Available online: https://www.bioenergyconsult.com/solid-waste-nigeria (accessed on 20 December 2018).
61. U.S. IEA Statistics. Ghana-Combustible Renewables and Waste (% of Total Energy). OECD/IEA. 2014. Available online: http://www.iea.org/stats/index.asp (accessed on 29 June 2021).
62. Deep Resource World Press (DRW). Energy Conversion Factors. 2012. Available online: https://deepresource.worldpress.com/2012/04/23/energyrelatedconversionfactors (accessed on 14 September 2018).
63. World Nuclear Association (WNA): Heat Values of Various Fuels. 2018. Available online: www.world-nuclear.org (accessed on 14 October 2018).
64. Ibadan Electricity Distribution Company (IBEDC). Kwara Needs 270 MW for 24-Hour Electricity. 2016. Available online: https://www.ilorin.info/fullnews.php?id=18082 (accessed on 3 October 2018).
65. Akinnuli, B.; Ogedengbe, T.I.; Oladosu, K.O. Computer Aided Design and Drafting of Helical gears. *J. Eng. Trends Appl. Sci.* **2012**, *3*, 959–968.

sustainability

Article

Processing Dross from Hot-Dip Galvanizing by Chlorination Roasting

Nurlan Kalievich Dosmukhamedov [1,*], Arkady Kaplan [2], Erzhan Esenbaiuly Zholdasbay [1], Gulzada Myngyshkyzy Koishina [1], Yeleussiz Bolatovich Tazhiev [1], Aidar Argyn [1], Yerzhan Itemenovich Kuldeyev [1] and Valery Kaplan [3]

[1] Department of Metallurgy and Mineral Processing, Satbayev University, Almaty 050000, Kazakhstan; zhte@mail.ru (E.E.Z.); gulzada.koishina@mail.ru (G.M.K.); eleusiz_t1990@mail.ru (Y.B.T.); aidarargyn@gmail.com (A.A.); e.kuldeyev@satbayev.university (Y.I.K.)
[2] Arvak Tech. LLC, Rehovot 76000-76878, Israel; arkadyc@gmail.com
[3] Weizmann Institute of Science, Rehovot 76000-76878, Israel; valery.kaplan@weizmann.ac.il
* Correspondence: nurdos@bk.ru

Abstract: Dross from hot-dip galvanizing is an important source of pure zinc ingots and zinc oxide for use as mineral additives in animal and poultry feed. Thermodynamic calculations have shown the possibility of solving the issue of dross processing by roasting using $CaCl_2$ and NH_4Cl. The influence of the consumption of chlorinating reagents, the roasting temperature on the degree of sublimation of Pb, Fe, Ni, Cu and Cd has been investigated. It has been shown that the best results are achieved when roasting the dross with the simultaneous use of $CaCl_2$ and NH_4Cl in amounts of 6 and 15% by weight of the feed material. The optimal roasting parameters were established: T = 1000 °C, duration—60 min, air flow—0.1 L/min. Recovered pure zinc oxide composition (%) was: 0.05 Pb, 0.15 Fe, 0.06 Ni, 0.003 Cu and 0.001 Cd. The degree of sublimation of copper, nickel and iron chlorides was ~75%, with lead and cadmium at 90–98% of their initial amount in the dross.

Keywords: dross; roasting; calcium chloride; ammonium chloride; impurities; sublimation; extraction; zinc; zinc oxide

Citation: Dosmukhamedov, N.K.; Kaplan, A.; Zholdasbay, E.E.; Koishina, G.M.; Tazhiev, Y.B.; Argyn, A.; Kuldeyev, Y.I.; Kaplan, V. Processing Dross from Hot-Dip Galvanizing by Chlorination Roasting. *Sustainability* **2021**, *13*, 12530. https://doi.org/10.3390/su132212530

Academic Editors: Rita Khanna, Yury Konyukhov and Igor Burmistrov

Received: 21 September 2021
Accepted: 2 November 2021
Published: 12 November 2021

1. Introduction

World zinc production is showing a slight increase, with consumption reaching ~14 million tons per year [1,2]. The main uses of zinc include the processes of galvanizing products. Hot-dip galvanized steel accounts for 50% of global zinc consumption [3]. High prices for zinc, which make up the main part of the cost of galvanizing, require its economical use and disposal of its waste, in particular, dross. The amount of dross during galvanizing is around 0.5 to 3.5% of the mass of the treated products. Despite the fact that, in terms of its composition, dross mainly consists of zinc oxide, 30–40% of metallic zinc remains in it [4,5]. Zinc ash represents 12 to 15% of the total amount of zinc used for piece hot galvanizing (leaching of zinc from zinc ash originating from hot-dip galvanizing) [5]. It is estimated that treatment of one ton of steel generates about 10 kg of the zinc dross and 9 kg of the zinc ash, giving a potential total zinc stream of 15–18 kg to recover [6]. Statistical data shows that every year in the European Union about 19 kg of steel per inhabitant is galvanized. For comparison, the average values calculated per inhabitant reach 12 kg in Poland, 26 kg in Belgium and 32 kg in Austria [7]. The Hot-dip Galvanized Steel Market report forecasts continuous increase in global production of galvanized steel at a compound annual growth rate of 5.1% in the period of 2019–2025 [8]. The complex chemical composition of the dross makes it difficult for further usage. Its main processing in practice is about the recovering of the metallic zinc.

Due to its poor quality, dross is not a marketable product and, in most cases, is reused in the process of galvanizing products. A significant content of lead and other impurity metals remain in the oxide part of the dross, so it has limited further applications. This problem is an important issue for enterprises in Kazakhstan, where steel products are galvanized. Due to the development of the construction industry, the demand for a wide range of galvanized steel products (pipes, channels, squares, wire, etc.) has sharply increased in the republic. The increase in the capacity of enterprises for the hot-dip galvanizing of steel products has led to large volumes of waste dross starting to accumulate, with storage occupying large areas. Further use and processing of dross is constrained by the increased content of Pb, Fe, Cu and other impurities in it, as well as the lack of rational processing technology.

Currently, in practice, pyrometallurgical methods of dross processing to obtain metallic zinc are widely used [9–15]. In recent years, hydrometallurgical methods have also been developing [1–5]. The most common way of dross processing is their heating in cylindrical drums or retorts of various designs to a temperature higher than the melting point of zinc [9]. The disadvantage of these methods is the lack of sealing the internal volume of the drum. The second significant drawback is the absence of direct contact of the walls of the heated combustion chamber with the raw material loaded into the drum.

The patented pyrometallurgical methods of dross processing with the use of various reagents (acids, alkaline methods, the addition of sodium and aluminum fluorides, etc.) have become widespread [10–15]. The main disadvantages of these works include: the laboriousness of the separation of the metal fraction; the formation of a large volume of wastewater requiring additional purification and evaporation to obtain zinc chloride; energy costs associated with the use of additional equipment; and an increase in the number of workers. One of the most effective methods for dross processing can be the preliminary separation of the metal and oxide parts of the dross, with further separate processing of each of them [16]. The metal part is melted in an induction furnace under a layer of ammonium chloride and charcoal. Metallic zinc is obtained from the metal part with the composition (%): 95.9 Zn, 1.54 Pb, 0.9 Fe and 0.4 Cu. The oxide part of the dross is subjected to roasting at 800–900 °C and zinc oxide is obtained, which is used for the preparation of whitewash. The elevated content of lead (1.2%) and iron (0.95%) in the oxide part of the dross does not allow obtaining high quality zinc oxide white paint.

Despite the indicated disadvantages, the approach proposed in [16] for the preliminary separation of the metal and oxide parts of the dross, with further separate processing of each of them, seems to be quite effective. At the same time, if the metal part can be used as a material for secondary use in galvanizing, then investigating the ways to process the non-metallic, oxide part of the dross containing such impurities as Pb, Fe, As and Sb requires additional research. In this regard, the use of metal chlorination has a great interest for the processing of dross. Characteristics of metal chlorides, such as their low melting point, high volatility and solubility in water, make it possible to recover valuable metals from various wastes in the form of their chlorides [17,18]. In terms of the complexity of the formation of metal chlorides and the difference in their properties, metals can be selectively chlorinated and sublimated by controlling the reaction temperature and vapor pressure of the products participating in the reaction [19–21].

In practice, chlorine gas, hydrogen chloride and alkali metal chlorides are used as chlorinating reagents. The use of $CaCl_2$, which has a high ability to chlorinate metals and has a rather low cost, has become widespread among them. This work presents the results of the reduction of impurity metals (Pb, Fe, etc.) from the dross by its roasting with $CaCl_2$ and NH_4Cl. The results of comprehensive investigation on the study of the influence of the consumption of chlorinating reagents and temperature on the behavior of Pb, Fe, Ni, Cu and Cd under roasting conditions are presented. Based on thermodynamic calculations of the reactions of the interaction of metal impurities with $CaCl_2$ and NH_4Cl, the possibility of obtaining pure metallic zinc and a mineral zinc additive (ZnO) suitable for use in animal and poultry feed is shown.

The importance of the obtained results is about their application for the development of a reductive-chlorinating technology aimed at utilization of accumulated substandard wastes of the hot-dip galvanizing process, for the additional production of commercial zinc metal and pure zinc oxide. The technology improves the quality of the environment by reducing harmful emissions into the environment (Pb, etc.), which fully meets the priority areas of scientific, technical and socioeconomic development of the Republic of Kazakhstan.

2. Experimental Part

- Experimental procedure

The initial sample of dross was crushed on a Cutting Mill SM 300 and subjected to sieve analysis with separation of the metal ($Zn_{met.}$) and oxide (ZnO) fractions. The metal fraction in the amount of 50 g was melted in a high-temperature shaft furnace (SiC heaters) at a heating rate of 10 °C/min with the addition of 3% NH_4Cl to the initial sample. Once the specified melting time had elapsed, the crucible with the material was removed from the furnace, cooled, and metallic zinc was separated.

The oxide fraction of the dross was pre-mixed with different consumption of chlorinating reagents ($CaCl_2$, NH_4Cl), loaded into an aluminum oxide boat and roasted in a horizontal furnace at a given temperature in an air flow. The gases from the furnace were directed to a gas-cleaning bottle for absorption. After the required holding time had elapsed, the furnace was cooled in a stream of air, after which the crucible with the obtained material (cinder) was removed from the furnace and broken. The cinder was weighed and analyzed for the content of impurity metals.

- Research methods

A SiC heating furnace was used as an experimental equipment for melting the metal fraction of dross. For roasting of oxide fraction of dross, Nabertherm RT 50–250/12 horizontal electric furnace with temperature controller B410 was used. For melting and for roasting, alumina crucibles and alumina boats (120 × 60 × 20 mm), respectively, were used as a container for loading the feed materials. The airflow rate during roasting of the oxide fraction of dross with $CaCl_2$ and NH_4Cl was monitored with RM-GC 0.04 rotameter.

Preparation of the samples for chemical analysis was carried out as follows: 10 mg of the material was dissolved in 65% (vol.) nitric acid and then analyzed. Material composition was characterized using an atomic absorption spectrophotometer equipped with a graphite combustion chamber (Perkin Elmer 5100). Powder X-ray diffraction (XRD) was performed on an Ultima III diffractometer (Rigaku Corporation, Japan) with quantitative phase analysis accomplished using Jade_10 (MDI, Cal.) software and the ICSD database and energy-dispersive X-ray fluorescence spectroscopy was performed on a LEO Supra scanning electron microscope (SEM).

Each sample obtained after the experiments was subjected to the elemental composition twice. The final elemental composition was determined based on the average value obtained from the results of two independent measurements. For a better understanding of the mechanism of roasting, a detailed thermodynamic analysis of the interaction of dross components with chlorinating reagents ($CaCl_2$ and NH_4Cl) was carried out.

- Feed material

The dross obtained after galvanizing steel pipes at the plant in Karaganda city was used as a feed material. The initial dross after grinding was subjected to sieve analysis and divided into classes. The results of the classification of the dross and the yields of the products of each class are shown in Table 1.

Three large fractions of dross are a metallic phase, consisting of zinc beads of different sizes with a small content of impurities. The results of XRD analysis and the phase composition of the fine fraction of the sample (−35 mesh) showed that it is mainly represented by the oxidized phase (ZnO) with small impregnations of metal in the form of a zinc–iron alloy, the surface of which is covered with a zinc oxide film (Figure 1). As the results of the phase composition of the fine oxidized dross fraction (Figure 1) show, it consists of 97.6% ZnO and 2.4% Zn–Fe alloy. Other impurity metals (Pb, Cu, Cd) are not displayed in the X-ray diffraction pattern due to the low content.

Table 1. Results of sieve analysis and product yield by class.

Size Fraction	+10 Mesh	−10 + 18 Mesh	−18 + 35 Mesh	−35 Mesh

wt.% of each size fraction	17.4%	8.9%	8.9%	64.9%

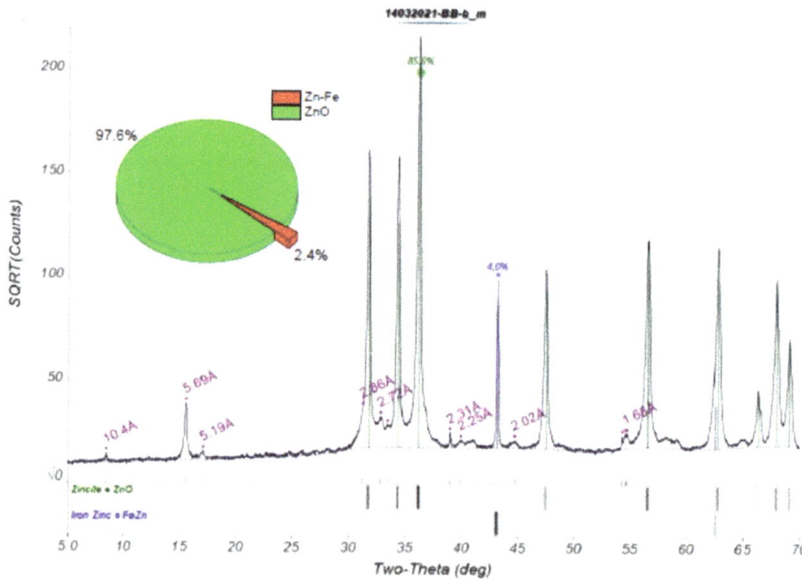

Figure 1. X-ray diffraction pattern and phase composition of the oxidized fraction.

From the results of SEM analyses, it can be seen (Figure 2) that, in the fine oxidized fraction, there are many particles of irregular shape. Two typical forms (point 1—dark area; point 2—pronounced white areas) can be seen: in the first corresponding area, the main components in the form of ZnO are represented; in the second area is an alloy of zinc and iron. Elemental EDS analysis shows a zinc content of 93.38%, which is in good agreement with the result of the phase composition established by the XRD method (Figure 1).

The results of the chemical analysis of the oxide fraction presented in Table 2 shows the presence of such impurities as Pb, Ni, Cu and Cd.

Table 2. Impurities content in the oxidized fraction.

Element	Pb	Fe	Ni	Cu	Cd
Content, wt.%	0.34	0.54	0.3	0.06	0.002

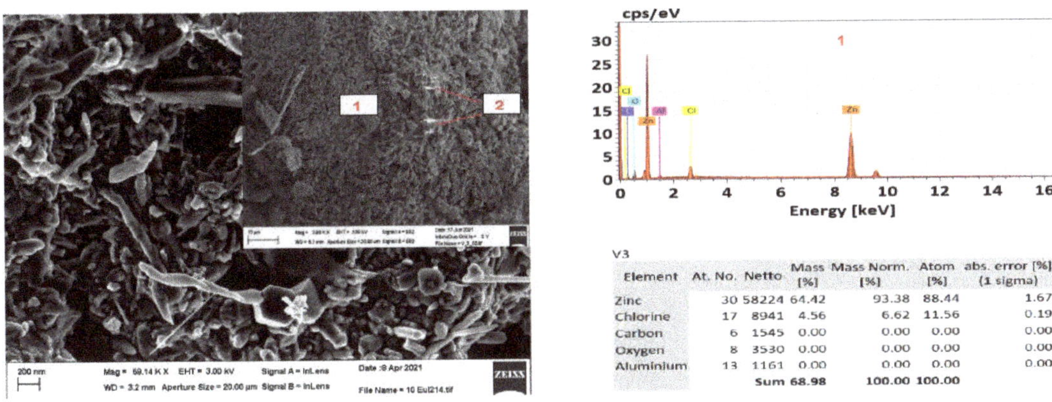

(a) (b)

Figure 2. SEM (**a**) and EDS (**b**) images of the oxidized fraction.

3. Results and Discussion

3.1. Thermodynamic Analysis of the Reactions of Chlorination of Zinc, Impurity Metals and Their Oxides with Calcium and Ammonium Chlorides

The model of thermodynamic reactions of chlorination of the components of the metal fraction with ammonium chloride (melting) and the oxidized fraction of dross (roasting) with calcium and ammonium chloride is considered from the standpoint of the above results of mineralogical studies of the phase composition of the initial products. The mechanism of the chlorination process can be described by a system of reactions typical for the conditions of melting the metal fraction of dross with ammonium chloride and roasting the oxidized fraction with $CaCl_2$ and NH_4Cl, shown in Table 3.

Table 3. Calculated values of the Gibbs free energy of the reactions of chlorination of zinc, impurity metals and their oxides with calcium and ammonium chlorides.

#	Reactions	Temperature, K							
		673	773	873	973	1073	1173	1273	1373
		ΔG°_T, kJ/mol							
(1)	$Zn + CaCl_2 + 0.5O_2 = ZnCl_2 + CaO$	−189.1	−180.6	−172.21	−164.07	−155.4	−144.8	−134.2	−123.7
(2)	$ZnCl_2 + 0.5O_2 = ZnO + Cl_2$	31.5	28.9	26.7	24.8	23.2	21.9	20.8	19.9
(3)	$Pb + CaCl_2 + 0.5O_{2(g)} = PbCl_2 + CaO$	−67.6	−74.5	−81.2	−87.9	−93.6	−97.2	−100.5	−103.6
(4)	$Fe + 1.5CaCl_2 + 0.75O_{2(g)} = FeCl_3 + 1.5\ CaO$	−65.4	−51.9	−38.5	−25.2	−10.6	7.3	25.4	43.7
(5)	$Cu + CaCl_2 + 0.5O_{2(g)} = CuCl_2 + CaO$	3.7	11.4	18.7	25.7	33.1	42.1	50.9	59.7
(6)	$Ni + CaCl_2 + 0.5O_{2(g)} = NiCl_2 + CaO$	−77	−67.03	−57.14	−47.32	−36.78	−24.21	−11.54	−2.96
(7)	$Cd + CaCl_2 + 0.5O_{2(g)} = CdCl_2 + CaO$	−160.51	−150.58	−142.33	−136.94	−131.1	−123.5	−116	−108.65
(8)	$ZnCl_2 + Pb + 0.5O_{2(g)} = PbCl_2 + ZnO$	−162.04	−166.86	−171.3	−175.4	−179.2	−182.7	−186	−189.2
(9)	$3ZnCl_2 + 2Fe + 1.5O_{2(g)} = 2FeCl_3 + 3ZnO$	−414.1	−380.91	−347.16	−312.95	−277.9	−242	−205.7	−136.26

Table 3. *Cont.*

#	Reactions	Temperature, K							
		673	773	873	973	1073	1173	1273	1373
		ΔG°_{T}, kJ/mol							
(10)	$ZnCl_2 + Cu + 0.5O_{2(g)} = CuCl_2 + ZnO$	−90.7	−80.9	−71.23	−61.73	−52.43	−43.36	−34.55	−25.84
(11)	$ZnCl_2 + Ni + 0.5O_{2(g)} = NiCl_2 + ZnO$	−171.42	−159.39	−147.18	−134.81	−122.3	−109.7	−97.09	−88.53
(12)	$ZnCl_2 + Cd + 0.5O_{2(g)} = CdCl_2 + ZnO$	−254.93	−242.93	−232.36	−224.44	−216.6	−209	−201.5	−194.23
(13)	$Zn + 2NH_4Cl + 0.5O_{2(g)} = ZnCl_2 + 2NH_{3(g)} + H_2O_{(g)}$	−249.52	−279.21	−309.16	−339.55	−370.3	−401.6	−433.2	−465.27
(14)	$ZnO + 2NH_4Cl = ZnCl_2 + 2NH_{3(g)} + H_2O_{(g)}$	−81.99	−133.48	−186.03	−239.54	−293.9	−349.1	−405	−461.67
(15)	$Pb + 2NH_4Cl + 0.5O_{2(g)} = PbCl_2 + 2NH_{3(g)} + H_2O_{(g)}$	−244.03	−300.35	−357.33	−414.94	−473.1	−531.8	−591.1	−650.87
(16)	$Fe + 3NH_4Cl + 0.75O_{2(g)} = FeCl_3 + 3NH_{3(g)} + 1.5H_2O_{(g)}$	−330.04	−390.68	−452.63	−515.79	−579.8	−644.6	−710.4	−777.13
(17)	$Cu + 2NH_4Cl + 0.5O_{2(g)} = CuCl_2 + 2NH_{3(g)} + H_2O_{(g)}$	−172.7	−214.39	−257.27	−301.28	−346.3	−392.4	−439.5	−487.51
(18)	$Ni + 2NH_4Cl + 0.5O_{2(g)} = NiCl_2 + 2NH_{3(g)} + H_2O_{(g)}$	−253.42	−292.88	−333.21	−374.36	−416.2	−458.8	−502.1	−550.21
(19)	$Cd + 2NH_4Cl + 0.5O_{2(g)} = CdCl_2 + 2NH_{3(g)} + H_2O_{(g)}$	−336.93	−376.42	−418.4	−463.9	−510.5	−558.1	−606.6	−655.9
(20)	$PbO + 2NH_4Cl = PbCl_2 + 2NH_{3(g)} + H_2O_{(g)}$	−157.17	−206.56	−259.76	−314.03	−369.2	−425	−479.6	−534.97
(21)	$Fe_2O_3 + 6NH_4Cl = 2FeCl_3 + 6NH_{3(g)} + 3H_2O_{(g)}$	−24.53	−173.18	−324.18	−477.23	−632.2	−789.2	−948.1	−1108.8
(22)	$CuO + 2NH_4Cl = CuCl_2 + 2NH_{3(g)} + H_2O_{(g)}$	−78.92	−129.27	−180.67	−233.07	−286.4	−340.6	−395.7	−451.75
(23)	$NiO + 2NH_4Cl = NiCl_2 + 2NH_{3(g)} + H_2O_{(g)}$	−74.82	−122.97	−171.92	−221.62	−272	−323	−374.7	−431.15
(24)	$CdO + 2NH_4Cl = CdCl_2 + 2NH_{3(g)} + H_2O_{(g)}$	−144.47	−194.53	−247	−302.99	−359.9	−417.7	−476.4	−535.83

The probable direction of the reactions was estimated by the change in the thermodynamic values of the system. Thermodynamic calculations were conducted by calculating the Gibbs free energy (ΔG°_{T}) of reactions and establishing their dependence on temperature. As the standard state of zinc, impurities metals and their oxides, the pure solid metal (Me_s.) and solid oxide (MeO_s.) was taken. For metal chlorides, the pure $MeCl_{2(g)}$ was considered. For the calculations, the reference data from [22] and the NIST-JANAF Thermochemical Tables website (http://kinetics.nist.gov/janaf, accessed on 20 September 2021) were used. The results of thermodynamic calculations are summarized in Table 3.

It was found that the course of interaction reactions between metal oxides MeO (Me-Zn, Pb, Fe, Cu, Ni, Cd) and $CaCl_2$ in the entire investigated temperature range is impossible due to the positive values of the Gibbs free energy. During the roasting, active chlorination of metallic zinc with calcium chloride according to reaction (1) is expected. In the temperature range 673–1373 K, the Gibbs free energy of reaction (1) shows a decrease with increasing temperature: from $\Delta G_{673\ K} = -189.1$ kJ/mol to $\Delta G_{1373\ K} = -123.7$ kJ/mol. Nevertheless, the high negative values of the Gibbs free energy indicate the possibility of the formation of gaseous zinc chloride and solid calcium oxide in the cinder over the entire investigated temperature range.

Zinc chloride formed as a result of reaction (1) does not interact with atmospheric oxygen according to reaction (2) due to the positive values of Gibbs free energy in the entire investigated temperature range of 673–1373 K. Thus, it can be argued that, under conditions of joint roasting of the oxide fraction of dross with $CaCl_2$ in an oxidizing atmosphere, favorable conditions are created for the conversion of metallic zinc into oxide. From a practical point of view, this will lead to an increase in the content of zinc oxide and, as a result, improve the quality of the final product (ZnO).

For the melting of the metal fraction, the use of calcium chloride does not seem to be effective due to the favorable conditions being created for the conversion of metallic zinc into its oxide. This will have a significant effect on reducing the quality of the resulting metal zinc ingot. The interaction of impurity metals with $CaCl_2$ is described by a group of reactions (3)–(7). At the temperature of melting (673 K) and roasting (1373 K), all reactions, except for reaction (5), are characterized by negative values of Gibbs free energy. High negative values of ΔG°_T are typical for reactions (3), (6) and (7), which indicates a deep sublimation of lead, nickel and cadmium, both during the melting of the metal fraction and during the roasting of the oxidized fraction.

Under roasting conditions, the possibility of the interaction of impurity metals with zinc chloride formed by reaction (1) is not excluded. The mechanism of the interaction of impurity metals with $ZnCl_2$ is described by the course of reactions (8)–(12). The results of thermodynamic calculations show large negative values of the Gibbs free energy of reactions (8)–(12). The high probability of reactions (8)–(12) indicates their priority over reactions (2), (3)–(7), which describe the direct chlorination of impurity metals with calcium chloride. The course of reactions (8)–(12) creates favorable conditions for the deep sublimation of impurity metals in the form of their chlorides, which significantly affects the improvement of the quality of the resulting zinc oxide.

The mechanism of roasting of oxide fraction of dross with ammonium chloride (Table 3) is described by a group of reactions (13)–(24). Large negative values of the Gibbs free energy of reactions (13)–(24) show a high probability of their occurrence and contribute to the production of purer zinc oxide due to the deep sublimation of impurity metals in the form of their chlorides. The values of the Gibbs free energy of reactions (13)–(24) are much higher than the values of ΔG°_T for reactions (3)–(7) (Table 3).

The obtained results indicate the preference and higher efficiency of melting the metal fraction of dross with ammonium chloride than with calcium chloride. The choice and justification of one or another chlorinating reagent for each specific process (melting, roasting) should be determined on the basis of the results of experimental tests, to study the effect of their consumption on the sublimation depth of impurity metals from both the metal fraction of the dross and the oxidized fraction.

3.2. Melting of Metal Fraction of Dross with Ammonium Chloride Addition

The initial sample in the amount of 50 g, prepared by mixing three large metal fractions of zinc (Table 1), was melted in a shaft silite furnace. Impurity content in the averaged sample: Pb—0.8%, Fe—0.22%, Cu—0.15%, Cd—0.005%. Previously, NH_4Cl was added to the initial sample in an amount of 2% of the sample weight. Melting was carried out at a temperature of 450 °C in an inert gas (Ar) atmosphere with a flow rate of 0.2 l/min. The duration of the experiments was 60 min.

After melting, an ingot of metallic zinc was obtained. The results of SEM and EDS analysis of the metal zinc ingot obtained after melting (Figure 3) showed the presence of exclusively zinc in it, which confirms its high quality.

3.3. Roasting of Oxidized Dross Fraction with $CaCl_2$ and NH_4Cl

The scheme of the installation for studying the influence of the consumption of chlorinating reagents ($CaCl_2$, NH_4Cl) and temperature on the production of commercial zinc oxide, suitable for use as mineral additives in animal and poultry feed, is shown in Figure 4.

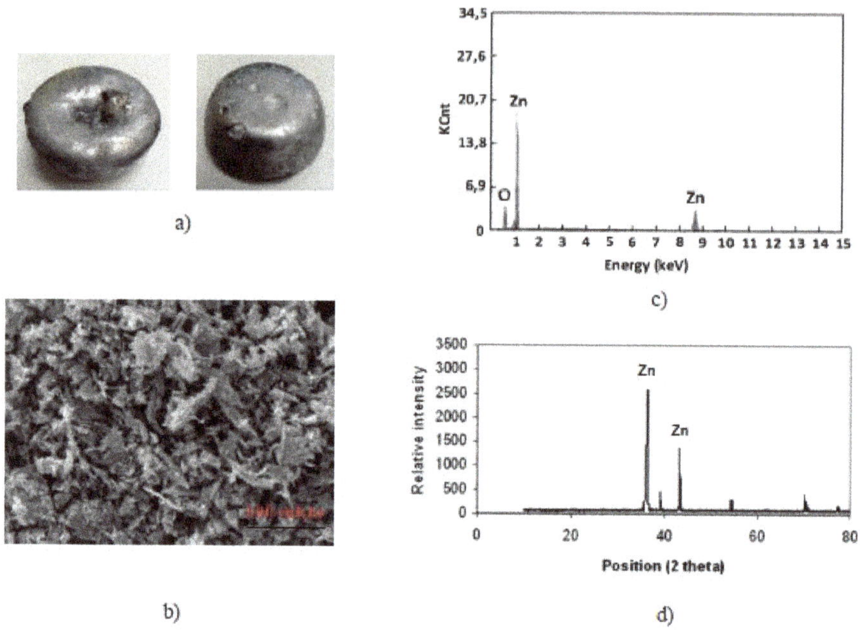

Figure 3. SEM (**b**), EDS (**c**) and XRD (**d**) results of metal zinc ingot (**a**).

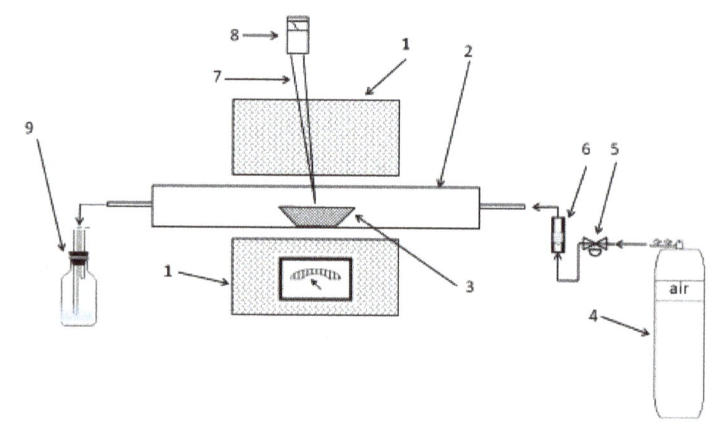

Figure 4. Laboratory setup scheme. 1—Nabertherm R50/250/12 electric furnace with temperature controller B410; 2—quartz reactor; 3—alundum boat with sample; 4—air cylinder; 5—valve; 6—rotameter RM-GC/0,04; 7—thermocouple type R; 8—secondary device KSP-4; 9—gas-cleaning bottle.

The experiments were carried out according to the method described above. The oxide fraction of the dross after hot galvanizing was mixed with a given consumption of calcium chloride and ammonium chloride in various mass ratios. The mixture was held in an alumina crucible inside the furnace, which was heated and then held at a predetermined temperature in an air flow that was supplied at a rate of 100 mL/min. Off-gases from the furnace were absorbed in a gas absorption vessel. After settling and cooling in a stream of air, the crucible was removed from the furnace and broken. The final product (cinder) was weighed and analyzed in accordance with the methods described above.

In the first series of experiments, the influence of the consumption of each separately taken chlorinating reagent ($CaCl_2$, NH_4Cl) on the degree of sublimation of impurities was studied. In all experiments, the temperature was constant and amounted to 1000 °C. The dependence of the final content of impurity metals in the cinder obtained after roasting on the consumption of $CaCl_2$ and NH_4Cl is shown in Figure 5.

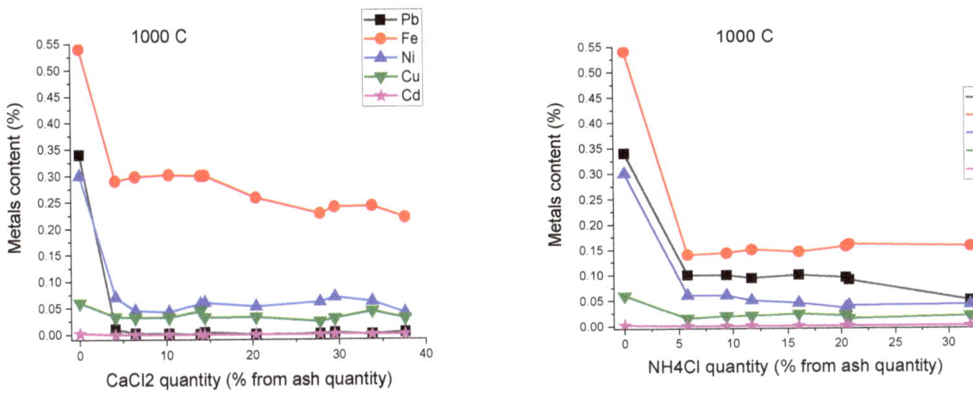

(**a**) (**b**)

Figure 5. Dependence of the content of metal impurities on the consumption of $CaCl_2$ (**a**) and NH_4Cl (**b**).

Comparative analysis of the results shown in Figure 5 indicates that an increase in the consumption of both $CaCl_2$ and NH_4Cl has a strong effect on reducing the content of all metal impurities in the final product (cinder). In the case of roasting the oxidized dross fraction jointly with $CaCl_2$, the effect of $CaCl_2$ consumption on the iron content is insignificant (Figure 5a). In the range of $CaCl_2$ consumption from 6 to 37% of the weight of the feed material, the iron content in the cinder decreases slightly, from 0.3 to 0.25%. At the same time, the content of lead and other impurity metals shows a sharp decrease, even at low consumption of $CaCl_2$. As can be seen in Figure 5a, the addition of $CaCl_2$ to the initial sample in an amount of 6% of its weight provides a minimum lead content of 0.002% in the final cinder. A further increase in the consumption of $CaCl_2$ has no effect on the decrease in the lead content in the cinder. This tendency is typical for copper, nickel and cadmium: at a $CaCl_2$ consumption equal to 6% of the weight of the feed material, the minimum contents of copper, nickel and cadmium in the cinder are achieved, which remain practically at the same level regardless of the increase in the $CaCl_2$ consumption.

Roasting the oxidized dross fraction with the addition of NH_4Cl provides a minimum iron content of up to 0.15% in the final cinder (Figure 5b). At the same time, the general character of the curve of the dependence of the content of impurity metals (Cu, Ni, Cd) on the NH_4Cl consumption (in the range of variation in the consumption from 5 to 37% of the weight of the initial material), repeats the course of the curve of a similar dependence established for impurity metals in the case of roasting using $CaCl_2$ (Figure 5a). Figure 5b shows that the consumption of NH_4Cl from 5 to 20% by weight of the feed material has no significant effect on the lead content in the resulting cinder. At low consumption of NH_4Cl (5%), although a sharp decrease in the lead content from 0.55 in the initial material to 0.12% is achieved, a further increase in the consumption has only a slight decrease in it in the cinder. It was found that the optimal consumption of NH_4Cl, providing a simultaneous minimum value of all metal impurities, corresponds to the consumption of NH_4Cl, equal to 15% of the weight of the feed material. This achieves a minimum lead content (0.1%) in the cinder.

The obtained results fully confirm the data of thermodynamic analysis and show that the use of $CaCl_2$ and NH_4Cl as a chlorinating reagent provides a deep purification of the initial material from such impurities as Pb, Cu, Ni and Cd. The best solution seems to be the simultaneous use of $CaCl_2$ and NH_4Cl. As the results of the experiments show, the optimal composition of the mixture, consisting of $CaCl_2$ and NH_4Cl, corresponds to consumption: 6% $CaCl_2$ and 15% NH_4Cl by weight of the feed material. Figure 6 shows the results of studying the effect of the roasting temperature of the oxidized dross fraction on the sublimation of impurity metals using a mixture of $CaCl_2$ and NH_4Cl with their consumption equal to 6 and 15% of the weight of the feed material, respectively.

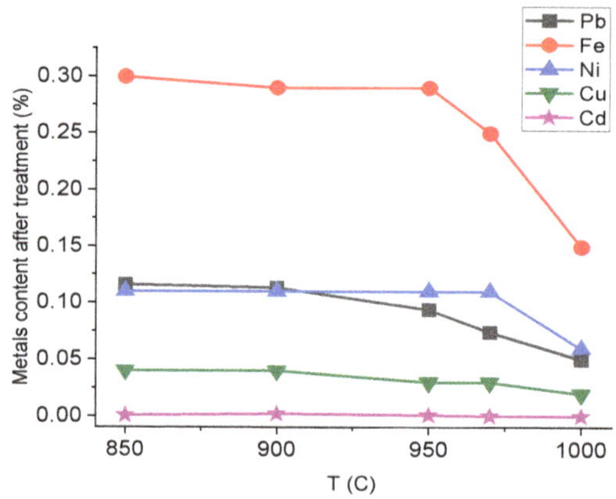

Figure 6. Dependence of the content of metal impurities on temperature. Consumption of $CaCl_2$ 6% and NH_4Cl 15%, by weight of the feed material.

It was found that the active removal of metal impurities from the cinder is observed in the temperature range 950–1000 °C. The best results were achieved at a roasting temperature of 1000 °C: the degree of sublimation of copper, nickel and iron chlorides is ~75% and lead and cadmium are 90–98% of their initial amount in the dross. Composition of the obtained cinder (wt.%) is 0.05 Pb, 0.15 Fe, 0.06 Ni, 0.003 Cu and 0.001 Cd, and fully complies with the requirements for pure zinc oxide used as mineral additives in animal and poultry feed.

4. Conclusions

This article presents a new method for obtaining pure metallic zinc and zinc oxide from hot-dip galvanizing dross by chlorination of a previously separated oxidized and non-metallic fraction. It was estimated that treatment of one ton of steel generates about 10 kg of the zinc dross and 9 kg of the zinc ash, giving a potential total zinc stream of 15–18 kg to recover. The results of thermodynamic calculations have shown that the use of $CaCl_2$ and NH_4Cl as chlorinating reagents makes it possible to obtain commercial metallic zinc and pure zinc oxide by deep reduction of impurity metals (Pb, Fe, Cu, Ni, Cd).

It was found that the greatest effect is achieved when melting the metal fraction with ammonium chloride. The mechanism of the process is described, and the optimal melting parameters are established, which ensure the production of a pure metal ingot of zinc: NH_4Cl consumption—2% by weight of the feed material, T = 450 °C, melting duration—60 min. The results were obtained on the influence of $CaCl_2$ and NH_4Cl consumption on the degree of sublimation of impurity metals under roasting conditions at a temperature of 1000 °C. It was found that the use of $CaCl_2$ as a chlorinating reagent provides deep purification of the feed material from impurities such as Pb, Cu, Ni and Cd. At the same time, the maximum reduction in the iron content is not possible. It is shown that the use of NH_4Cl as a chlorinating reagent, although it provides deep removal of impurity metals, does not succeed in reducing the lead content: the lead content in the cinder remains at the level of 0.12%.

The best solution seems to be the simultaneous use of $CaCl_2$ and NH_4Cl. As the results of the experiments show, the optimal composition of the mixture, consisting of $CaCl_2$ and NH_4Cl, should correspond to the consumption of 6% $CaCl_2$ and 15% NH_4Cl by weight of the feed material. The optimal parameters of roasting were established as: T = 1000 °C, duration—60 min, air consumption—0.1 l/min. Technological indicators of the process: composition of the obtained pure zinc oxide (wt%): 0.05 Pb, 0.15 Fe, 0.06 Ni, 0.003 Cu and 0.001 Cd; the degree of sublimation of chlorides of copper, nickel and iron ~75% and lead and cadmium 90–98% of their initial amount in the dross.

Author Contributions: Conceptualization, N.K.D. and V.K.; methodology, E.E.Z.; software, G.M.K.; validation, A.K., Y.B.T. and A.A.; formal analysis, Y.I.K.; investigation, A.A.; resources, N.K.D.; data curation, G.M.K.; writing—original draft preparation, A.A.; writing—review and editing, V.K.; visualization, N.K; supervision, G.M.K.; project administration, N.K.D.; funding acquisition, G.M.K. All authors have read and agreed to the published version of the manuscript.

Funding: This study received funding from the Science Committee of the Ministry of Education and Science of the Republic of Kazakhstan, project No. AP09058297 "Development of a new waste-free technology for the disposal of hot-dip galvanizing waste with complex extraction of valuable components".

Institutional Review Board Statement: Not applicable.

Informed Consent Statement: Not applicable.

Data Availability Statement: The references were presented in the text of the manuscript.

Acknowledgments: The research was carried out within the framework of grant funding of the Science Committee of the Ministry of Education and Science of the Republic of Kazakhstan for 2021–2023 in the priority area "Rational use of natural resources, including water resources, geology, processing, new materials and technologies, safe products and structures" project No. AP09058297, "Development of a new waste-free technology for the disposal of hot-dip galvanized wastes with complex extraction of valuable components".

Conflicts of Interest: On behalf of all authors, the corresponding author declares that there is no conflict of interest.

References

1. Saramak, D.; Krawczykowski, D.; Gawenda, T. Investigations of zinc recovery from metallurgical waste. *IOP Conf. Ser.: Mater. Sci. Eng.* **2018**, *427*, 012017. [CrossRef]
2. Trpcevska, J.; Rudnik, E.; Holkova, B.; Laubertova, M. Leaching of Zinc Ash with Hydrochloric Acid Solutions. *Pol. J. Environ. Stud.* **2018**, *27*, 1771–1785. [CrossRef]
3. Stubbe, G.; Hillmann, C.; Wolf, C. Zinc and Iron Recovery from Filter Dust by Melt Bath Injection into an Induction Furnace. *Erzmetall* **2016**, *69*, 5–12.
4. Konstantinov, V.M.; Gegenya, D.V.; Bogdanchik, M.I. Market overview for zinc and zinc waste. In *Foundry Processes*; MSTU: Russia, Moskow, 2014; 293p.
5. Takácová, Z.; Hluchánová, B.; Trpcevská, J. Leaching of zinc from zinc ash originating from hot-dip galvanizing. *Metall* **2010**, *64*, 517–519.
6. Schmitz, D.; Friedrich, B. In-house recycling of hard zinc and zinc ash by liquid metal centrifugation. In *Proceedings of the EMC*; GDMB: Düsseldorf, Germany, 2007; pp. 1–20.
7. Smakowski, T.; Galos, K.; Lewicka, E. *Balance of Management of Mineral Resources of Poland and the World 2013*; Panstwowy Instytut Geologiczny: Warszawa, Poland, 2013.
8. QYR Steel Research Center. *Global Hot-Dip Galvanized Steel Market Insights, Forecast to 2025*; QYR Steel Research Center: Beijing, China, 2019. Available online: https://www.reportsnreports.com/reports/1821020-global-hot-dip-galvanized-steel-market-insights-forecast-to-2025.html (accessed on 1 September 2021).
9. Najiba, S. Recovery of Zinc from Ash of Galvanizing Plant by Hydrometallurgical Route. Master's Thesis, Bangladesh University of Engineering and Technology, Dhaka, Bangladesh, 2009; pp. 1–92.
10. Yagubova, V.L.; Chumaevsky, O.V. Method for Producing Zinc Nitrate. RF Patent No. 96111199/25, 4 October 1998.
11. Barkhatov, V.I.; Dobrovolsky, I.P.; Kapkaev, Y.S.; Kostyunin, S.V. Method for Processing Waste Containing Heavy Non-Ferrous Metals. RF Patent No. 2016108776, 26 May 2017.
12. Kazantsev, G.F.; Barbin, N.M.; Moiseev, G.K.; Vatolin, N.A. Method for Processing Zinc Waste. RF Patent No. 99107789/02, 4 October 2000.

13. Chernov, P.P.; Koryshev, A.N.; Larin, Y.I. Method for Producing Zinc from Zinc Dross. RF Patent No. 2001109810/02, 27 August 2002.
14. Yudin, R.A.; Vinogradov, A.V.; Kovryakov, S.V.; Sudakov, E.A.; Yanichev, A.N. Installation and Method of Extracting Zinc from Zinc Ash. RF Patent No. 2008102795/02, 10 October 2009.
15. Kodochigov, B.N. Method for Processing Zinc Ash. RF Patent No. 2267546, 10 January 2006.
16. Tarasov, A.V. Recycling of Hot-Dip Galvanized Waste. *Steel* **1989**, *6*, 57–58.
17. Wang, H.; Feng, Y.; Li, H.; Kang, J. The separation of gold and vanadium in carbonaceous gold ore by one-step roasting method. *Powder Technol.* **2019**, *355*, 191. [CrossRef]
18. Wang, H.; Feng, Y.; Li, H.; Kang, J. Simultaneous extraction of gold and zinc from refractory carbonaceous gold ore by chlorination roasting process. *Trans. Nonferrous Met. Soc. China* **2020**, *30*, 1111. [CrossRef]
19. Guo, X.; Zhang, B.; Wang, Q.; Li, Z.; Tian, Q. Recovery of Zinc and Lead from Copper Smelting Slags by Chlorination Roasting. *JOM* **2021**, *73*, 1861–1870. [CrossRef]
20. Bai, S.; Bi, Y.; Ding, Z.; Li, C.; Wen, S. Innovative methodology for the utilization of low-grade pyrite cinder containing heavy metals via hydrothermal alkali melting followed by chlorination roasting. *J. Alloys Compd.* **2020**, *840*, 155722. [CrossRef]
21. Qin, H.; Guo, X.; Tian, Q.; Zhang, L. Pyrite enhanced chlorination roasting and its efficacy in gold and silver recovery from gold tailing. *Sep. Purif. Technol.* **2020**, *250*, 117168. [CrossRef]
22. Turkdogan, E.T. *Physical Chemistry of High Temperature Technology*; Academic Press: New York, NY, USA, 1980; 462p.

sustainability

MDPI

Review

Red Mud as a Secondary Resource of Low-Grade Iron: A Global Perspective

Rita Khanna [1,*], Yuri Konyukhov [2], Dmitry Zinoveev [3], Kalidoss Jayasankar [4], Igor Burmistrov [5], Maksim Kravchenko [6] and Partha S. Mukherjee [7]

1 School of Materials Science and Engineering (Ret.), The University of New South Wales, Sydney, NSW 2052, Australia
2 Department of Functional Nanosystems and High-Temperature Materials, National University of Science and Technology "MISiS", 119049 Moscow, Russia; ykonukhov@misis.ru
3 A.A. Baikov Institute of Metallurgy and Materials Science, Russian Academy of Science, 119334 Moscow, Russia; dzinoveev@imet.ac.ru
4 Materials Science & Technology Division, CSIR-National Institute for Interdisciplinary Science and Technology, Thiruvananthapuram 695019, India; jayasankar.met@gmail.com
5 Engineering Centre, Plekhanov Russian University of Economics, 117997 Moscow, Russia; burmistov.in@rea.ru
6 Moscow Power Engineering Institute, National Research University, 111250 Moscow, Russia; Kravchenkomv@mpei.ru
7 Institute of Minerals and Materials Technology (Ret.), Council of Scientific and Industrial Research, Bhubaneshwar 751013, India; psmukherjee52@gmail.com
* Correspondence: rita.khanna66@gmail.com; Tel.: +61-043415-5956

Abstract: Managing red mud (RM), a solid waste byproduct of the alumina recovery process, is a serious ecological and environmental issue. With ~150 million tons/year of RM being generated globally, nearly 4.6 billion tons of RM are presently stored in vast waste reserves. RM can be a valuable resource of metals, minor elements, and rare earth elements. The suitability of RM as a low-grade iron resource was assessed in this study. The utilization of RM as a material resource in several commercial, industrial operations was briefly reviewed. Key features of iron recovery techniques, such as magnetic separation, carbothermal reduction, smelting reduction, acid leaching, and hydrothermal techniques were presented. RMs from different parts of the globe including India, China, Greece, Italy, France, and Russia were examined for their iron recovery potential. Data on RM composition, iron recovery, techniques, and yields was presented. The composition range of RMs examined were: Fe_2O_3: 28.3–63.2 wt.%; Al_2O_3: 6.9–26.53 wt.%; SiO_2: 2.3–22.0 wt.%; Na_2O: 0.27–13.44 wt.%; CaO: 0.26–23.8 wt.%; Al_2O_3/SiO_2: 0.3–4.6. Even with a high alumina content and high Al_2O_3/SiO_2 ratios, it was possible to recover iron in all cases, showing the significant potential of RM as a secondary resource of low-grade iron.

Keywords: red mud; iron recovery; waste utilization; smelting; low grade iron ore; reduction

Citation: Khanna, R.; Konyukhov, Y.; Zinoveev, D.; Jayasankar, K.; Burmistrov, I.; Kravchenko, M.; Mukherjee, P.S. Red Mud as a Secondary Resource of Low-Grade Iron: A Global Perspective. *Sustainability* **2022**, *14*, 1258. https://doi.org/10.3390/su14031258

Academic Editors: Castorina Silva Vieira and Marc A. Rosen

Received: 10 December 2021
Accepted: 20 January 2022
Published: 23 January 2022

1. Introduction

Bauxite, recognized as the key aluminum ore, contains ~30–54% alumina (Al_2O_3) along with mixtures of silica, iron oxides, titanium dioxide, and several impurities, etc. Up to 95% Al is produced worldwide from the bauxite ore using the Bayer's process. In the Bayer process, bauxite is washed in a hot solution of sodium hydroxide, which leaches Al in the form of $Al(OH)_3$; it is later calcined to form Al_2O_3. Red mud (RM), also known as the bauxite residue, is a solid waste byproduct of the alumina recovery process. Producing one ton of alumina consumes up to 2–3 tons of bauxite and generates ~0.4–2 tons of RM depending on the source/location of the ore; the global average of RM is ~1.3 tons per ton of alumina [1,2]. Alumina has also been extracted from poor-quality diasporic bauxite using soda lime sintering at high temperatures [3]. Wang and Liu (2012) [4] have reported on the chemical composition of Bayer and sintering red muds from an alumina refining plant in Guizhou, China [4]. While the typical composition of Bayer RM was Fe_2O_3 (26.4%), Al_2O_3 (18.9%), SiO_2 (8.52%), CaO (21.8%), and TiO_2 (7.4%), the corresponding results for the sintering RM were Fe_2O_3 (7.9%), Al_2O_3 (17.3%), SiO_2 (17.3%), CaO (40.2%), and TiO_2 (7.4%). The sintering RM showed much higher levels of CaO and SiO_2 but low levels of Fe_2O_3 as compared to the Bayer RM; similar trends have also been reported by Feng and Yang (2018) [5]. Due to quite low levels of Fe_2O_3 present, sintering RM is not considered a potential iron resource in this study.

Global bauxite resources are estimated to be ~55 to 75 billion tons with reserves in Africa (32%), Oceania (23%), South America, the Caribbean (21%), and Asia (18%). With the annual primary aluminum production increasing to 47 million tons, an estimated 150 million tons of RM are being produced in various aluminum plants across the world [6,7]. Nearly 4.6 billion tons of RM, presently stored around in vast waste reserves, represents massive industrial waste scenarios [8–10]. Several red mud incidents in different countries have also been tabulated in a number of publications [11,12].

Extensive efforts and attempts have been made to recycle, process, and utilize RM waste. Several excellent reviews are available in the literature on RM recycling, utilization, and management [13–20]. Key challenges in processing RM include, among others, high alkalinity pH: 10–13 [21], high moisture content, small particle sizes ($\leq$75 μm), large volumes, and transportation costs [22]. Landfilling and dumping around the industrial plants have been the standard procedure; vast quantities of stockpiled RM are known to be toxic and very hazardous [23]. Deep sea dumping and storage in ponds have also been attempted. Poorly designed storage dams are likely to fail under certain circumstances resulting in local and environmental contamination [24]. Numerous efforts are being made to find economically viable and environmentally sustainable solutions to the RM problem. However, most current options can only accommodate a small fraction of the red mud generated globally [25,26].

Depending on the region and the location, the type of bauxite, operating parameters of the Bayer process, and the composition of red mud can vary over a wide range [27]. Major constituents of RM are: Fe_2O_3: 26.6–46 wt.%; Al_2O_3: 15–21.2 wt.%; TiO_2: 4.9–21.2 wt.%; SiO_2: 4.4–18.8 wt.%; Na_2O: 1–10.3 wt.%; CaO: 1–22.2 wt.% [28,29]. Several minor elements (U, Ga, V, Zr, Sc, Cr, Mn, Y, Ni, Zn, Th, rare earths) may also be present in ppm levels [30]. RM particles are generally fine-grained with particle sizes smaller than 75 μm for up to 90 wt.% of particles, and surface areas ranging between 10–30 m^2/g [31]. Given the complex composition of RM, fine particle sizes and toxic caustic-corrosive nature, the processing of RM can be quite challenging, and technically difficult [32].

Global steel production has more than tripled over the past 50 years; China and India are the top two steel producing nations. In 2020, the global steel production was 1.86 billion tons; China accounted for more than a billion tons of steel [33]. Iron ore resources are getting consumed at a fast rate leading to a reduced availability of high-grade iron ores as well as sharp declines in their supply to iron and steel plants. The current scenario is steadily shifting towards the use of low grade iron ores. Presently, red mud is not considered as competitive raw material for iron and steel making as the Fe_2O_3 content of the red mud (27–47%) is significantly lower than the requisite concentrations in conventional iron ores (>60%) [34]. Iron ores with Fe contents above 65% are regarded as high-grade ores; 62–64% medium grade ores and those below 58% Fe are considered as low-grade ores [35,36]. The known world resources of crude iron ores are approximately 800 billion tons [37]. Most of the known deposits contain low-grade ores with iron contents even less than 30%. In order to meet the growing demand for iron and steel, it is thus imperative to find new sources of iron ore to supplement the existing reserves [38].

With significant reserves of iron, albeit in relatively low concentrations, the RM waste could be considered as a low-grade iron ore. However, requisite high volumes, fine particle sizes, high moisture, and alkaline content make it virtually impossible and too costly to transport large quantities of RM to distant places. Commercial scale iron recovery, therefore, needs to be carried out near the source of RM generation. Across the globe, there are wide variations in RM compositions as well as methodologies used to recover iron.

The aim of this study is to develop a critical review and an assessment on the recovery of iron from RMs in different parts of globe. This study will consolidate relevant data from India, China, Greece, Italy, France, and Russia in the form of case studies to determine the likelihood of iron recovery processes, tailored to local conditions and RM compositions. The article is organized as follows. A brief overview on RM waste management strategies is presented in Section 2. Section 3 presents key techniques used for the recovery of iron from red mud. Consolidated results from the global scenario are presented in Section 4 in the form of case studies. These discussions are followed by a comparison of RMs with low grade iron ores and concluding remarks in Sections 5 and 6.

2. RM Waste Management Strategies: A Brief Overview

A brief overview of RM waste management practices is presented next with specific focus on recycling, utilization, and material recovery. Attempts have been made to recycle large volumes of RM in several commercial, industrial operations such as building and construction materials, cement, concrete, coloring agents for paints, paper/polymer/ceramic/refractory products, catalysts, inorganic chemicals, adsorbents, metallurgical recovery of metals (Al, Ti, Si), rare earth elements, and others [39,40]. Being the focus of this study, specific details on the recovery of iron from red mud will be presented in later sections. Some of the key approaches to RM utilization are summarized below.

2.1. Construction Materials

2.1.1. Cement

The use of RM in the production of cement has been investigated extensively in several countries. Road construction using RM is considered one of the most suitable applications [41]. High contents of iron and aluminum in RM were reported to be beneficial for cement production, accelerating the clinkering process and for slag replacement. Pontikes and Angelopoulos [42] have recommended the addition of ~30 kg of RM per ton of ordinary Portland cement. Feng et al. [43] mixed 50 wt.% thermally active RM with modifiers along with other solid wastes to produce Portland cement; the flexural and compressive strength of the manufactured concrete were found to be adequate for concrete requirements of pavement materials. Liu and Poon [44] used RM to prepare self-compacting concrete; Nikbin et al. [45] evaluated the mechanical strength, elastic moduli and environmental impact of lightweight concrete. RM from HINDALCO (Hindustan Aluminum Corporation, Mumbai, India) has been used to prepare superior strength Portland cements; typical compositions used were: 30–35% RM, 15–20% bauxite, 7.5–10% gypsum, 45–50% lime [46]. The crushing and bending strength of the mortar were found to improve with 5–10% RM addition [47].

2.1.2. Building Materials

Dodoo-Arhin et al. [48] investigated composites of RM (20–50 wt.%) and Tetegbu clay (balance) for making ceramic bricks. Sintered bricks were fired in the temperature range 800–1100 °C; optimal properties for the construction brick applications were obtained for RM contents of 50 wt.%. Several researchers [49–54] have examined the potential recycling of RM for construction and building supplies, different types of bricks such as non-fired, non-steam cured bricks, ceramic glazed tiles, decorative bricks, hollow bricks, fly ash bricks, and others; their strength and quality was found to be quite comparable to standard bricks. Stabilized blocks with strength of grades II/III bricks have been developed mixing RM with clay and/or fly ash; low density (1.1–1.2 g/cm^3), hollow bricks and blocks have also been prepared [55,56].

2.1.3. Glass-Ceramics, Geopolymers, Catalysts

Some other applications of RM include utilization in the production of glass-ceramics, geopolymers, catalysts, and fillers, among others. Firing blends of RM with limestone at 1150 to 1200 °C, Amritphale et al. [57] were successful in producing cements with a strength of 200 kg/cm^2. Ceramic tiles were also produced from mixtures of RM with fly ash/additives; relative proportions of various constituents were used for controlling the tile strength [58,59]; RM has also been used to reduce the impact of glazing [60]. Yang et al. [61] were able to prepare a CaO-SiO_2-Al_2O_3 glass-ceramics using up to 85% waste materials (RM and fly ash). The limitations of RM in terms of colour and impact strength and densification mechanism could limit some commercial applications [62].

Geopolymers and synthetic aluminosilicates have been produced from mixtures of RM with fly ash other wastes such as FeNi slag, rice husk ash, blast furnace slag, etc. as a replacement for Portland cement ceramic and composite applications [63,64]. Advanced composites are used extensively in defence, aerospace, petrochemical, and marine industries [65]. Using RM as base and filler materials, a 4 km long highway was trialed in Zibo, Shandong Province, China and was found to meet requisite strength requirements [66]. RM has been used as a catalyst in several applications, e.g., hydro-dechlorination, coal hydrogenation, coal/biomass liquefaction, SO_2 reduction, methane combustion etc. [67,68]; However, only a small fraction of RM waste can be used in these applications. Environmental pollution control applications of RM include waste water treatment, soil remediation with bauxite residues, waste gas purification and sulphur removal, etc. [69–71].

2.2. Resource Recovery of Metals

Depending on the region & the location, the type of bauxite, operating parameters of the Bayer process, the composition of red mud can vary over a wide range [72]. Major constituents of RM are: Fe_2O_3 Al_2O_3, TiO_2, SiO_2, Na_2O, CaO etc. [73–76]. Several minor elements (U, Ga, V, Zr, Sc, Cr, Mn, Y, Ni, Zn, Th, rare earths) may also be present in ppm levels [77–79]. RM particles are generally fine-grained with particle sizes smaller than 75 µm for up to 90 wt.% of particles, and surface areas ranging between 10–30 m^2/g. Given the complex composition of RM, fine particle sizes, high moisture content and toxic caustic-corrosive nature, the processing of RM can be quite challenging and technically difficult [32,80]. Some of the techniques used to extract key metals from RM are briefly summarized below.

2.2.1. Aluminium

Up to 4–16% aluminum could be present in RM after Bayer's process as residual alumina; several hydro/bio metallurgical approaches have been developed for processing RM as an additional alumina resource. Vachon et al. [81] used mixtures of sulphuric/citrus/oxalic acids in a range of proportions for extracting aluminum from RM. These authors also investigated bioleaching using acid-producing bacteria and fungi recovering up to 75% Al present. Bruckard et al. [82] used fluxing, smelting and leaching at 60 °C with soda and calcium carbonate to dissolve up to 55% Al. The hydrometallurgical approach involves atmospheric/pressurized alkali leaching at high temperatures as well as acid leaching. Zhang et al. [83,84] were able to recover alumina by forming andradite–grossular hydrogarnet in a hydrothermal process. The pyrometallurgical recovery involves sintering with soda (25–70 wt.%) and lime (10–50 wt.%) in the temperature range 1000–1050 °C to form water-soluble sodium aluminate, followed by leaching in water/NaOH (1.5–2 M) achieving up to 76–90% aluminum extraction [85,86]. However, some of these processing methods could be quite expensive, time consuming and produce secondary waste products as well.

2.2.2. Titanium

Pyrometallurgical as well as hydrometallurgical techniques have been used for extracting TiO_2 from RM. Calcining RM in the temperature range 850–1300 °C, followed by smelting with a carbonaceous reductant in an electric furnace, produces molten iron along with a slag containing titanium dioxide, alumina, silica, etc. [87–89]. Kasliwal and Sai [90] leached RM with HCl at 363 K, and heat treated the leach residue with Na_2CO_3 at 1423 K for 115 min to recover up to 76% TiO_2. Deep et al. [91] used hydrolysis and calcination followed by acid leaching with (3% H_2O_2, 0.5 M H_2SO_4, 4 M HCl/H_2SO_4) solution to recover TiO_2. The commercialization of these predominantly laboratory based techniques would require significant resources and associated infrastructure.

2.2.3. Rare Earths and Other Elements

RM also contains several rare earth elements (REEs), including scandium (60–120 ppm), yttrium (60–150 ppm), gallium (60–80 ppm), as well as radioactive elements uranium (50–60 ppm) and thorium (20–30 ppm) [92–97]; some of these concentrations are comparable or even better than corresponding concentrations in respective ores. REEs are generally recovered from RM using the hydrometallurgical route. Wang et al. [98] recovered scandium from synthetic RM using solvent extraction with Di-(2-ethylhexyl) phosphoric acid. Through selective leaching of a Greek RM with dilute HNO_3, Ochsenkuhn-Petropulu et al. [99] were able to recover up to 90% Y, ~70% heavy lanthanides (Yb, Dy, Er), ~50% middle lanthanides (Nd, Sm, Eu, Gd) and 30% light lanthanides (La, Ce, Pr). Mineral as well as organic acids were used by Ujaczki et al. [100] to leach out REEs from the RM. Selective recovery of up to 75% Sc can be achieved from RM by using the sulfation method [101]

3. Iron Recovery from Red Mud: Key Techniques

Iron is one of the key constituents of RM and is often present as oxides or oxyhydroxides. Several techniques such as magnetic separation, carbothermal reduction, microwave carbothermal reduction, smelting reduction, acid leaching, and hydrothermal techniques have been developed to separate and recover iron-bearing phases [102,103]. A brief summary of the methods used is presented below.

3.1. *Magnetic Separation*

The main mineral phase of iron in RM tailings could be hematite, goethite, or magnetite, which are magnetic or weakly magnetic materials. One of the most effective ways to separate the fine magnetic particles in liquid suspensions is through direct application of high intensity magnetic fields; electromagnets with fields up to 0.3 T (Tesla) have been used for recovering Fe from RM slurry. The yield was generally quite poor with the resulting output in the form of magnetic and non-magnetic fractions [104]. A direct magnetic separation approach is inefficient for Fe recovery from RM due to the lack of magnetic iron-bearing mineralogical phases. Much stronger fields (up to 4 T) using superconducting magnets have also been used to separate the fine magnetic particles (<100 μm) in liquid suspensions and to separate iron rich and iron poor fractions [105]. After the separation process, high iron content and low iron content parts of RM could be recycled as resources in different industrial sectors. Higher magnetic intensity leads to more amounts of concentrates being collected and higher iron contents in the concentrates. Using hydro cyclone and magnetic separation techniques, it was possible to raise the iron content of RM from 52 wt.% to 70 wt.% [106].

3.2. *Carbothermal Reduction*

The carbothermal reduction of red mud has been reported using several techniques such as smelting [107], microwave reduction [108], suspension roasting [109], and sintering [110]. In the carbothermal reduction, RM powders are mixed with a range of carbonaceous materials (coals, coke, charcoal, graphite, bagasse, spent pot lining etc. [111,112] and sodium flux to produce pellets/briquets followed by heat treatment in the temperature range between 600–1200 °C for time periods between 30–240 min. This technique is limited by the low grade of iron produced, economic costs of the flux and reductants used, and high energy consumption.

3.3. Microwave Carbothermal Reduction

This technique uses isothermal/non-isothermal reduction using microwave heating followed by magnetic separation [113,114]. Microwave heating has several advantages over conventional heating, such as the instantaneous generation of heat inside moderately absorptive materials rather than relatively slow heat transfer from the outside surface. Typical processing conditions are 600–1000 W power, 700–1000 °C temperature range, 5–20 min roasting time and 5–16 wt.% C. Reductants used include charcoal and CO/CO_2 gases. Microwave heat treatment is followed by low intensity (0.1–1.8 T) magnetic separation. This treatment has significantly faster reaction times and potentially lower energy consumption as compared to conventional heat treatment methods [115]. Some of the limitations of this technique include high costs of reductants, low grade iron recovery and scaling issues.

3.4. Smelting Reduction

Smelting reduction involves the roasting of red mud, reductant and flux pellets in the temperature range 1400–1750 °C for times between 15 to 60 min. Furnaces used include resistance furnace, electric arc furnace, plasma furnace, vertical tube furnace, etc. [116–123]. Coal, coke, graphite, blast furnace sludge, and biomass have been used as reductants [124]. Additives during pellet making include Na_2CO_3, $CaCO_3$, Na_2SO_4-Na_2CO_3, among others. Under the ENEXAL project, Balomenos et al. [125] have reported on the reductive smelting of red mud and coke in electric arc furnace based experiments, preliminary thermodynamic modelling, and transition to full-scale pilot plant operation. Typical post treatments included the separation of metal and slag and acid leaching; end products were in the form of pig iron, iron nuggets, and REE enriched solutions. While high grade iron (91–96%) could be recovered, the limitations of this technique include high energy and flux consumption, high levels of phosphorus, and sulphur in iron and generally low yields.

3.5. Suspension Reduction

In this technique, RM undergoes reduction with reducing gases CO, CO_2, and H_2 combined with N_2 in a fluidized/static bed reactors [126–128]. This technique offers an alternate processing route for the separation of iron bearing phases from RM and the residue valorization via hydrogen reduction followed by magnetic separation. Typical temperature ranges and heating times were 350–600 °C for 30 min and 500–640 °C for 15–30 min. Key advantages of this method are relatively low processing temperatures, low energy consumption, no need for sintering or agglomerating the residue, and negligible carbon contamination typically observed in carbothermal reduction. This method is, however, limited by the high cost of H_2, poor iron enrichment, complex design, and composition variables.

3.6. Hydrometallurgy

A number of hydrometallurgical processes based on acid leaching have been developed for extracting iron from red mud. The metallic values in red mud are leached using mineral acids: HCl, H_2SO_4, HNO_3, H_3PO_4, and organic oxalic acid followed by solvent extraction [129]. An extraction rate of 96% could be achieved after leaching with 1 mol/L oxalic acid at 75 °C for 2 h. UV radiation was used on the precipitates for transforming ferric oxalate to ferrous oxalate [130]. Leaching with 8 N sulphuric acid at 100 °C for 24 h, an iron recovery of 47 wt.% could be achieved [131]. Iron grade recovery ranged from 94.2–97.4 wt.% for oxalic acid, 52–95.9 wt.% for hydrochloric acid, and 40–97.4 wt.% for sulphuric acid [132,133]. However, this technique is uneconomical for large scale utilization and nonselective iron extraction, and suffers from secondary waste generation and requires further downstream processing.

3.7. Other Techniques

Being ecological and requiring low energy consumption, bio-metallurgical techniques based on bioleaching are becoming increasingly important for the treatment of iron oxide ores [134,135]. Desulfuromonas palmitatis is believed to dissolve part of iron minerals from bauxite; especially 95% of amorphous ferrihydrite can be dissolved. The high pH value of red mud is likely to restrict the use of bioleaching for extracting of iron oxides [136]. Vakilchap et al. [137] observed that Aspergillus niger fungus can recover up to 69.8% Al, 60% Ti, and 25.4% Fe from red mud; a holding time of 30 days was required. Penicillium tricolor fungus has also been used for REE recovery from red mud [138,139]. In the hydrothermal method, red mud is digested in an alkaline medium at high temperatures (40% NaOH, 250 °C, 30–180 min), followed by washing and magnetic separation. [140]. Iron values present in red mud can be extracted using both the pyrometallurgical and hydrometallurgical techniques. Oxalic acid leaching, coupled with photocatalytic reduction, may be suitable for low iron grade red mud. Carbothermal microwave reduction followed by magnetic separation presents a high efficiency technique for red muds with 25–40% iron values. The choice of recovery technique will be based on economic feasibility, efficiency, as well as recovery efficiency.

4. Global Scenario

In this section, we present an overview on red muds being used a secondary iron resource in different countries around the globe. Details will be provided on the composition of red mud, iron recovery technique used, product yield and other relevant features in the form of case studies. As high temperature processing and pyrometallurgy are traditional approaches in conventional ironmaking, in this section we have focused our specific attention on the pyrometallurgical recovery of iron from red muds using well-established techniques.

4.1. INDIA

4.1.1. Case Study 1

In this study, the red mud was sourced from HINDALCO India and the Bauxite type was Laterite [113]. The chemical composition of RM is given in Table 1.

Table 1. Chemical composition of red mud [113].

Constituent	Fe_2O_3	Al_2O_3	SiO_2	TiO_2	CaO	Na_2O	Iron Grade
wt.%	36.26	16.58	8.32	17.10	1.43	6.00	25.7%

As received RM powder was pulverized using a ball mill to sizes below 100 μm and concentrated to an iron rich fraction using magnetic separation, followed by reduction roasting pre-treatment. Reduction investigations were carried out in the temperature range (700–1000 °C) using a silica crucible in a standard muffle furnace as well as in a 900 W microwave furnace. Charcoal was used as the reductant in both furnaces and the heating times were between 10–50 min. A coating of calcium oxide was used from trapping CO_2 gas over period of time. Specimens were water quenched after the heat treatments. Magnetic separation was carried out on reduced samples using 0.2 T magnetic field. Heat treatment temperature and heating time were found to be key factors in iron recovery. A treatment at 1000 °C for 10 min resulted in 95% iron recovery and an increase in RM iron content to 48.5 wt.%.

4.1.2. Case Study 2

In this study, the red mud was sourced from NALCO India; Bauxite type used was Laterite [114]. The chemical composition of RM is given in Table 2.

Experimental steps used in this study were quite similar to those in Case Study 1 and involved the blending of red mud with charcoal reductant, carbothermal reduction in muffle, and microwave furnaces followed by magnetic separation. Optimal experimental conditions for iron grade and iron recovery were determined. In the muffle furnace, optimal conditions for iron grade and iron recovery were 1000 °C, 16.5 wt.% C, 50 min and 850 °C, 11 wt.% C, 50 min, respectively. Corresponding conditions for the microwave furnace were 1000 °C, 11 wt.% C, 10 min and 850 °C, 5.5 wt.% C, 30 min, respectively.

Table 2. Chemical composition of red mud [114].

Constituent	Fe_2O_3	Al_2O_3	SiO_2	TiO_2	CaO	Na_2O	Iron Grade
wt.%	47.85	22.64	10.25	3.58	1.86	10.25	33.5%

4.1.3. Case Study 3

Jayasankar et al. [117] used thermal plasma technology to produce pig iron from RM, as well as from low grade iron ore fines. The RM was sourced from NALCO India; its composition is provided in Table 3. The term LOI represents 'loss on ignition'.

Table 3. Chemical composition of red mud [117].

Constituent	Fe_2O_3	Al_2O_3	SiO_2	TiO_2	CaO	Na_2O	LOI
wt.%	49.65	6.86	22.0	-	-	8.29	11.39%

The plasma furnace used in this study uses a pot type 35 kW DC extended transfer arc plasma reactor; zircon coated graphite crucibles were used as the furnace hearth [141]. Size reduced powders were placed in the graphite crucible of the plasma reactor. 1 kg of RM was mixed with 30% petroleum coke and 10% dolomite; the presence of dolomite helped in the separation of Fe from the slag. Graphite was also used as a reductant in some cases. These blends were loaded in the crucible. The plasma furnace was operated using 250 A current and a DC voltage of 60 V. Argon gas was passed continuously through the furnace at a flow rate of 1 L/min. Blends were smelted in the temperature range 1550–1650 °C for times between 20–30 min. The smelted pig iron and slag were collected into a preheated graphite mould through a tap hole. Iron recovery of 78% and 48% was achieved for graphite and petroleum coke reductants.

4.2. CHINA

4.2.1. Case Study 4

Li et al. [142] collected red mud from Shandong Province, China; the sample was pre-dried at 100 °C for 10 h to remove moisture. The chemical composition is given in Table 4.

Table 4. Chemical composition of red mud [142].

Constituent	Fe_2O_3	Al_2O_3	SiO_2	TiO_2	CaO	Na_2O	LOI
wt.%	47.76	10.26	2.25	4.22	0.26	1.13	11.93%

The RM, flux, and bentonite binder were mixed thoroughly to prepare green balls (800 mm dia., 200 mm thickness) in a disk pelletizer; these were later dried using a vacuum oven. Bituminous coals were used reductants and mixed with dry pellets with C/Fe mass ratio of 1.5. These were heat treated in a muffle furnace at 1050 °C for 80 min and furnace cooled. About 50 gm of pre-reduced pellets were mixed with 10% C and smelted at 1600 °C for 30 min; a corundum crucible was used during smelting. Pig iron and slag were cooled to room temperature under N_2 atmosphere for subsequent separation. The pig iron was produced with an iron recovery of 98.15% and the slag containing 43.17% Al_2O_3 and 15.71% TiO_2, respectively.

4.2.2. Case Study 5

Chun et al. [143] investigated the recovery of iron from red mud using high temperature reduction of carbon bearing RM briquettes. The red mud was collected from Shandong

Aluminum Corporation, China. Up to 80% particles could pass through a 0.1 mm sieve. The composition of red mud is given in Table 5.

Table 5. Chemical composition of red mud [143].

Constituent	Fe_2O_3	Al_2O_3	SiO_2	TiO_2	CaO	Na_2O
wt.%	42.45	10.35	6.89	-	0.72	1.36

The coal with the following characteristics was used as the reductant: fixed carbon—78.99%, volatile matter—7.4%, ash content—10.95%; sodium borate was used as an additive. Red mud, sodium borate (0–6 wt.%) and coal fines (7–12 wt.%) were blended using a small amount of water. Cylindrical briquettes were prepared in a mould by applying 8 MPa pressure for 1 min; these were then oven dried at 105 °C for 4 h. The reduction of briquettes was carried out in the temperature range 1200–1350 °C four times between 20 to 30 min in N_2 (1 L/min) atmosphere. Optimal roasting conditions were determined to be 1300 °C for 30 min. Reduced briquettes were ground finely, and the iron content was separated using magnetic separation (0.08 T). The addition of sodium borate was found to increase the iron grade of the recovered metallic powder to ~90%; which was attributed to the iron oxide reduction and the growth of metallic grains leading to an improved separation from gangue during subsequent magnetic processing.

4.2.3. Case Study 6

Zhang et al. [144] reported a semi-smelting reduction on an iron rich bauxite ore and magnetic separation process for the recovery of iron and alumina slag under a lower basicity condition. The composition of iron rich bauxite from western Guangxi, China is given in Table 6.

Table 6. Chemical composition of red mud [144].

Constituent	Fe_2O_3	Al_2O_3	SiO_2	TiO_2	CaO	Na_2O	LOI
wt.%	40.62	26.53	11.77	1.57	1.38	-	16.42%

Iron rich bauxite, $Ca(OH)_2$, anthracite coal, and CaF_2 were mixed thoroughly in a range of proportions and crushed using a ball mill. The anthracite coal with 78.73% fixed carbon was from Yangquan in Shanxi province in China. The average particle size and specific surface area of blends were determined to be 88.431 μm and 0.149 m^2/g. The mixtures were placed in a roller press and squeezed in the form of briquettes; these were then dried at 200 °C for 3 h. The composite briquettes were placed in a corundum crucible and reduced to a semi-smelting stage in a high temperature furnace in Ar (1.5 L/min) atmosphere. The reduction process was repeated three times. The reduced briquettes were crushed by the jaw crusher to 1–10 mm sizes and iron nugget and alumina slag were magnetically separated by a high strength magnetic field (0.4 T). Increasing the temperature can significantly increase the iron recovery and total iron content in the magnetic fraction; it also has a strong influence on the slag melting point and metallization rate of the briquettes. Reduction studies were carried out in the temperature range 1375–1450 °C for 1 to 15 min. Optimum conditions were determined to be: $w(CaO)/w(SiO_2)$ ratio of 1.25; 1425 °C, 15 min, anthracite proportion of ~12%. High quality iron nugget and high-grade alumina slag could be achieved with highest iron recovery rate (96.84 wt.%) and grade of alumina slag (43.98 wt.%), respectively.

4.2.4. Case Study 7

Guo et al. [145] used red mud from a Chinese aluminum plant and used magnetic separation to concentrate into iron rich fraction. The chemical composition of the iron rich RM is given in Table 7.

Table 7. Chemical composition of red mud [145].

Constituent	Fe_2O_3	Al_2O_3	SiO_2	TiO_2	CaO	Na_2O
wt.%	63.2	8.96	8.54	-	0.551	2.5

Red mud, anthracite coal, and lime-stone were used as raw materials; these were baked in a drying cabinet with forced convection at 200 °C for 2 h. These were mixed in a range of proportion to ensure sufficient carbon reductant and desired basicity. The mixture was homogenized with a binder and compacted into pellets using a roller press. Direct reduction of pellets was carried out in a muffle furnace using following operating conditions: 1350 °C/20 min; 1375 °C/25 min; and 1400 °C/30 min. Heat treated pellets were quenched in liquid N_2. The degree of metallization was determined using x-ray diffraction and chemical analysis. Temperature was found to be a key factor for reduction and the separation of metal and slag. Iron nuggets with an iron content of 96.52 wt.% were observed within the pellets after 1400 °C/30 min heat treatment. A chemical analysis of nuggets showed that the total iron was higher than in pig iron along with lower Si and Mn contents.

4.3. GREECE

4.3.1. Case Study 8

Samouhos et al. [146] carried out investigations on red mud supplied by the Greek aluminum producer "Aluminum of Greece S.A". The RM was relatively finely grained with 60 wt.% of particles with sizes below 100 μm and 97 wt.% with particles below 400 μm. The chemical composition of the RM is given in Table 8.

Table 8. Chemical composition of red mud [146].

Constituent	Fe_2O_3	Al_2O_3	SiO_2	TiO_2	CaO	Na_2O	LOI
wt.%	43.59	18.45	6.0	5.54	11.38	1.82	11.7%

This study investigated the separation of the iron oxide from RM using low temperature reduction by hydrogen under static conditions, followed by wet magnetic separation. 20 gm of dried red mud sample (105 °C overnight) was placed in an alumina crucible and positioned in the outer region of a horizontal furnace (room temperature). The furnace tube was continuously purged with N_2 gas (250 mL/min) to remove all oxygen present. The sample was then transferred to the central hot zone and the gas flow switched from N_2 to N_2/H_2 gas mixtures in a range of proportions. RM samples were reduced in the temperature range 300–450 °C for 30 min and were furnace cooled by withdrawing to the outer region. The conversion degree of hematite to magnetite was investigated as a function of reduction time, temperature, and H_2 supply. A maximum conversion of 87% was achieved at 480 °C. Reduced RM samples were subjected to magnetic separation yielding a magnetic fraction with over 54 wt.% magnetite content.

4.3.2. Case Study 9

Cardenia et al. [147] carried out investigations on the RM provided by Mytilineos, Metallurgy Business Unit (formerly known as AoG). The particle size distribution of the dried sample had a mean particle size (D_{50}) of 1.87 μm and D_{90} of 90% of the particles below 43 μm. The composition of red mud is given in Table 9.

Table 9. Chemical composition of red mud [147].

Constituent	Fe_2O_3	Al_2O_3	SiO_2	TiO_2	CaO	Na_2O	LOI
wt.%	43.51	19.25	6.5	5.49	9.58	2.80	9.41%

Dried homogenized bauxite residue was mixed with Na_2CO_3 in the ratio 4:1 and sintered in a muffle furnace at 900 °C for 2 h and furnace cooled. The sinter was stirred in a

0.1 M NaOH solution at 80 °C for 4 h with a 1.5% *w*/*v* pulp density and 240 rpm speed. Sodium and aluminum were leached out and a solid residue was left behind as modified red mud. The dried modified red mud was mixed with metallurgical coke in the ratio 4:9 and formed into pellets. These were roasted in a microwave at 600 W power, 300 s, and a N_2 gas flow of 1 L/min for three times. During the 1st and 2nd MW reductive roasting process, the resulting cinders were found to be rich in spherical metallic Fe particles trapped within the matrix. These were separated using a manual sieve. These were then subjected to a wet high intensity magnetic separation. A high concentration of metallic iron was detected in the magnetic fraction (56 wt.%) along with calcium, silicon, and titanium also present in minor quantities. Up to 4% iron was detected in the non-magnetic fraction. With successive MW roastings, the recovery of iron was further enhanced up to 69%.

4.3.3. Case Study 10

Borra et al. [107] used the red mud provided by the Aluminum of Greece Company (Agios Nikolaos, Greece). The composition of RM is provided in Table 10.

Table 10. Chemical composition of red mud [107].

Constituent	Fe_2O_3	Al_2O_3	SiO_2	TiO_2	CaO	Na_2O
wt.%	44.6	23.6	10.2	5.7	11.2	2.5

The RM sample was mixed with graphite (5–15 wt.%) and flux using a pestle and mortar; wollastonite ($CaSiO_3$) was used as a flux and graphite as a reducing agent. Handmade pellets were dried at 105 °C for 12 h. Smelting reduction experiments were carried out in a high-temperature vertical alumina tube furnace in the temperature range 1500–1600 °C with a heating rate 5 °C/min and kept at the pre-set temperature for 1 h. High purity Ar gas was flushed through the furnace at a flow rate of 0.4 L/min. After the heat treatment, the furnace was cooled down to room temperature. Optimum conditions at 1500 °C were identified as 20 wt.% wollastonite and 5 wt.% of graphite. More than 85 wt.% iron could be separated in the form of a nugget. Additional 10% iron could be recovered through subsequent grinding of slag and magnetic separation.

4.4. *ITALY*

Case Study 11

Mombelli et al. [124] used red muds from EurAllumina S.p.A., Portovesme (Italy) as a raw material for the production of iron-based secondary raw material in the form of cast iron and direct reduced iron (DRI). The composition of fresh and calcined red mud is given in Table 11. Red muds were calcined in a muffle furnace at 1000 °C for 60 min to eliminate hydroxide and carbonate species. Graphite and blast furnace sludge were used as reducing agents; lime was added as a fluxing agent to aid slag formation.

Table 11. Chemical composition of red mud [124].

Constituent (wt.%)	Fe_2O_3	Al_2O_3	SiO_2	TiO_2	CaO	Na_2O	LOI
Fresh RM	20.89	16.3	9.8	5.2	5.54	10.39	29.78
Calcined RM	28.26	21.05	15.69	7.26	7.69	13.44	3.78

A muffle furnace was used to carry out reduction experiments at 1200 °C, 1300 °C, and 1500 °C. The furnace containing the samples was heated to these temperature, maintained for 15 min, and then cooled. The results showed that the blast furnace sludge was a suitable reductant to recover the iron fraction in the red mud. The metallization degree achieved was higher than 70%; the composition 1:1 red mud/blast furnace sludges were found to be equivalent to 0.85 C/Fe_2O_3. Mombelli et al. [148] mixed RM with 6 and 12 wt.% metallurgical coke and slag conditioners CaO and SiO_2 and reduced the blend in an arc transferred plasma (ATP) reactor at 1600–1700 °C for 30–35 min. The quality of cast iron quality produced was evaluated for sake of valorization and marketability.

4.5. FRANCE

Case Study 12

Maihatchi et al. [149] have used an alternative route to beneficiate RM by directly producing electrolytic iron from hematite present in red muds in a highly alkaline medium. Red mud specimens were obtained from Alteo, a European alumina manufacturer. The chemical composition of RM is provided in Table 12.

Table 12. Chemical composition of red mud [149].

Constituent	Fe_2O_3	Al_2O_3	SiO_2	TiO_2	CaO	Na_2O
wt.%	52.7	13.9	4.19	7.4	4.1	2.1

A double-walled Pyrex cell with a total volume of 600 mL and diameter 80 mm was used in the experiments. Approximately 450 mL of 12.5 mol/L NaOH solution and 150 g of red mud were introduced into the cell. A cylindrical graphite rod with a 19.4 cm^2 wetted area was used as the working electrode. A platinum-coated titanium grid was used as a counter electrode. A saturated Ag/AgCl electrode was used as the reference electrode. A SP-150 potentiostat coupled to a 10 A booster acted as the current source. The solid: liquid ratio and amounts of impurities present in red mud were varied to optimize the faradaic yield and the production rate of electrolytic iron. Hematite could be reduced to iron with an efficiency over 80% for current densities up to 1000 A/m^2. Highest current efficiency for red muds were achieved for current densities below 50 A/m^2; these efficiencies were seen to decrease to 20% at 1000 A/m^2. The produced electrodeposit contained more than 97% metal iron in all cases. For the economic and commercial viability of this technique, it is requisite to obtain larger faradaic yields for the beneficiation process.

4.6. RUSSIA

4.6.1. Case Study 13

Valeev et al. [119] carried out reductive smelting on neutralized RM from Ural Aluminium Plant (Kamensk-Uralsky, Russia) with an aim to obtain the maximum amount of pig iron and to achieve a complete separation of the metal and slag. The chemical composition of RM is given in Table 13.

Table 13. Chemical composition of red mud [119].

Constituent	Fe_2O_3	Al_2O_3	SiO_2	TiO_2	CaO	Na_2O	LOI
wt.%	36.9	11.8	8.71	3.54	23.8	0.27	12.46%

The mixture with 27.6 g RM and 2.4 g graphite was loaded in a carbon crucible and placed in a Tammann furnace for reductive smelting. The furnace was heated up to 1300 °C rapidly for 1 h, then a slow heating to the required temperature range 1650–1750 °C at 10 °C/min. The assembly was kept at these temperatures for 10 min. The furnace was then switched off and the crucibles were furnace cooled to room temperature. After 1650 °C smelting reduction, the slag phase containing up to 8% Fe was not separated from the metal. Increasing temperature to 1700 °C saw the iron content in the slag to reduce to ~3% and the formation of large metal ingots at the crucible bottom. Almost complete separation of metal and slag was observed at 1750 °C with the iron content in the slag dropping to 0.16%. Pig iron obtained at 1750 °C could find application for producing moulds for steel-casting.

4.6.2. Case Study 14

Grudinsky et al. [150] have reported on the solid-state carbothermic roasting of two RMs in the presence of special additives followed by magnetic separation. Red muds were obtained from Bogoslovsky Aluminum Plant (Russian Federation, labelled as 'B') and from the Ural Aluminum Plant (Russian Federation, labelled as 'U'). The composition of 'B' and 'U' RMs are provided in Tables 13 and 14 respectively.

Table 14. Chemical composition of red mud [150].

Constituent	Fe_2O_3	Al_2O_3	SiO_2	TiO_2	CaO	Na_2O
wt.%	49.81	12.77	8.71	4.67	9.26	33

Long-flaming coal with 15% ash content and 18% moisture content was used as a reductant; Na_2SO_4 was used as an additive. RM, coal, and the additive were finely ground and blended together. These blends were roasted in a muffle furnace in the temperature range 1000–1450 °C, for times between 20–180 min. Heat treated samples were removed from the furnace and quenched with liquid N_2. Roasted RM samples were ground finely, and the iron rich fraction was magnetically separated using a magnetic field of 0.35 T. Iron concentrates with more than 95% of iron recovery and 83% of the iron grade were obtained for 'U' and 'B' red muds respectively. Optimal conditions for the roasting of the samples were a duration of 180 min with the addition of 13.65% Na_2SO_4 at 1150 °C and 1350 °C for the 'B' and the 'U' RMs respectively.

5. Comparison with Low Grade Iron Ores

The overall quality of an iron ore is judged mainly by its Fe content in the elemental form. Iron ores with Fe contents above 65% are considered high-grade ores [35,151]. Medium grade ores contain ~62–64% iron and those below 58% iron are low grade ores. Other important considerations include the concentrations of SiO_2, Al_2O_3, and S and P. The alumina to silica ratio higher than one can cause serious operational problems during sintering and subsequent smelting in the blast furnace [152]. High alumina content in the iron ore/and sinter leads to a viscous slag during smelting in a blast furnace resulting in a high consumption of coke as well as other concerns during tapping. Generalized concentrations of SiO_2 in iron ore should be less than 6 wt.% [153] and Al_2O_3 should be between 3–4 wt.%. [154] for blast furnace ironmaking.

In Figure 1, the Fe_2O_3, Al_2O_3, and SiO_2 contents of RMs from the 14 case studies have been plotted. The Fe_2O_3 content in most cases was between 30–50 wt.%; the average value was determined to be ~44 wt.%. The Al_2O_3 content of various RMs was quite high ranging between 6.9 to 26.5 wt.%; the SiO_2 content ranged between 2.3 to 22 wt.%.

In Figure 2, Al_2O_3/SiO_2 ratio vs. Fe_2O_3 content has been plotted for various red muds. Except for one RM, this ratio was higher than 1 (range: 1.3 to 4.6) for most RMs under investigation. Low Al_2O_3 in the iron ore facilitates and helps in maintaining lower alumina in slag without increasing slag rate and keeping it fluid. However, in sponge iron making (DRI) the SiO_2/Al_2O_3 ratio should preferably be 1. Higher SiO_2 in ore causes accretion formation in the DRI kiln; a balanced Al_2O_3 content enables easy dislodging of formed accretions. Some operations in China currently run on very high alumina iron ore, but the silica content is usually 3–4 times that of alumina.

China is currently world's largest producer of iron and produced 1053 Mt out of total global production of 1864 Mt [33,155]. Facing the risks of raw material shortages, China is working towards the feasibility of using domestic low grade/refractory iron ore resources. There is a large deposit in the Guangxi Zhuang Autonomous Region with an average of 43 wt.% Fe_2O_3 and 25 wt.% Al_2O_3 [156]. Rath et al. [157] have reported on a low grade iron ore from India containing 51.6 wt.% Fe_2O_3, 17.6 wt.% SiO_2 and 4.3 wt.% Al_2O_3.

In this study, iron recovery from a wide variety of red muds across the globe was reviewed in the form of case studies. Alumina content was higher than 20 wt.% in four case studies. In the case study 2 (Fe_2O_3: 47.9 wt.%; Al_2O_3: 22.64 wt.%; SiO_2: 10.25 wt.%; Al_2O_3/SiO_2: 1.92), the carbothermal reduction was carried out in muffle as well as microwave furnaces. Optimal conditions for iron recovery were determined as: muffle furnace: (1000 °C, 16.5 wt.% C, 50 min; 850 °C, 11 wt.% C, 50 min) and microwave furnace: (1000 °C, 11 wt.% C, 10 min and 850 °C, 5.5 wt.% C, 30 min). In case study 6 (Fe_2O_3: 40.6 wt.%; Al_2O_3: 26.53 wt.%; SiO_2: 11.77 wt.%; Al_2O_3/SiO_2: 2.25), high quality iron nuggets could be recovered with high iron recovery rate (96.84 wt.%) using semi-smelting reduction and magnetic separation process. In case study 10 (Fe_2O_3: 44.6 wt.%; Al_2O_3: 23.6 wt.%; SiO_2:

10.2 wt.%; Al_2O_3/SiO_2: 2.31), the smelting reduction experiments were carried out in the temperature range 1500–1600 °C; more than 85 wt.% iron could be recovered in the form of nuggets. In case study 11 (Fe_2O_3: 28.3 wt.%; Al_2O_3: 21.1 wt.%; SiO_2: 15.7 wt.%; Al_2O_3/SiO_2: 1.34), RM was mixed with 6 and 12 wt.% metallurgical coke and reduced in an arc transferred plasma (ATP) reactor at 1600–1700 °C for 30–35 min to achieve a high degree of metallization.

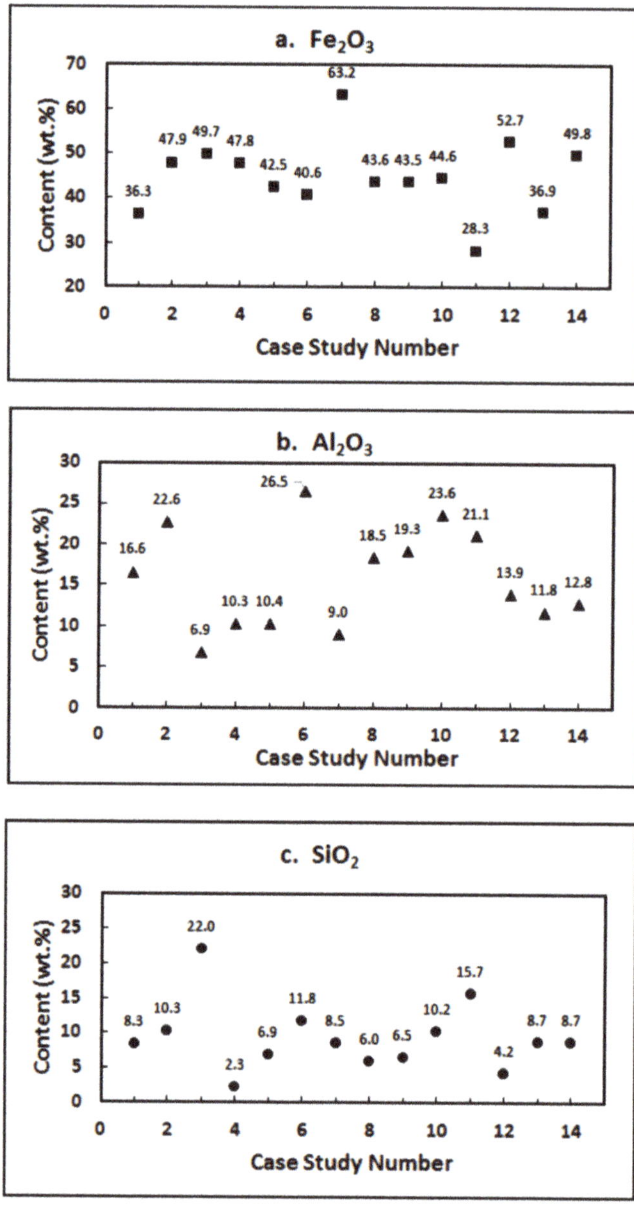

Figure 1. Concentrations of Fe_2O_3, Al_2O_3 and SiO_2 in red muds from case studies 1–14.

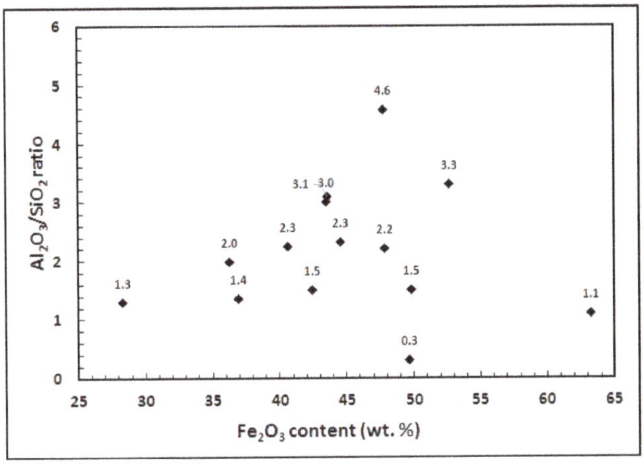

Figure 2. Al_2O_3/SiO_2 ratio vs. Fe_2O_3 concentration for red muds in case studies 1–14.

6. Concluding Remarks

In this article, we have focused our attention on evaluating the iron resource potential of RM. No attempt has been made towards establishing policies for managing RM waste or measures for a circular economy. Iron ore resources are getting consumed at a fast rate leading to a reduced availability of high-grade iron ores; the current scenario is steadily shifting towards the use of low-grade iron ores. Key findings from this study indicate that even with high levels of Al_2O_3 and SiO_2, red muds have a good potential to be used as a low-grade iron resource.

While RMs may not qualify as suitable blast furnace feed, these can find application as iron resource in DRIs, iron nuggets, sinters, etc. Alternatively, the blending of ores and waste products could play a significant role in the selection process. An economic and sustainable iron recovery operations may require more than a single feed source. Such an approach has been successfully used to partially replace coal/coke with waste carbonaceous products in iron and steelmaking [158,159]. Along with supplementing the natural resource sector, waste reutilization will enhance the environmental sustainability of materials sector, resource conservation, as well as reducing the carbon footprint of the industry.

Author Contributions: Conceptualization, original draft: R.K. and Y.K.; Supervision and resources: D.Z., K.J. and I.B. Formal analysis: M.K. and P.S.M. All authors have read and agreed to the published version of the manuscript.

Funding: This research received no external funding.

Institutional Review Board Statement: Not applicable.

Informed Consent Statement: Not applicable.

Conflicts of Interest: The authors declare no conflict of interest.

References

1. Zhang, R.; Zheng, S.; Ma, S.; Zhang, Y. Recovery of alumina and alkali in Bayer red mud by the formation of andradite-grossular hydrogarnet in hydrothermal process. *J. Hazard. Mater.* **2011**, *189*, 827–835. [CrossRef]
2. Archambo, M.; Kawatra, S.K. Red mud: Fundamentals and new avenues for utilization. *Miner. Process. Extr. Metall. Rev.* **2020**, *42*, 427–450. [CrossRef]
3. Yu, H.; Pan, X.; Dong, K.; Zhang, W.; Bi, S. The Sintering and Leaching of Low-Grade Diasporic Bauxite by the Improved Lime-Sintering Process. *Adv. Mater. Res.* **2013**, *616–618*, 1051–1054. [CrossRef]
4. Wang, P.; Liu, D.Y. Physical and Chemical Properties of Sintering Red Mud and Bayer Red Mud and the Implications for Beneficial Utilization. *Materials* **2012**, *5*, 1800–1810. [CrossRef]

5. Feng, Y.; Yang, C. Analysis on Physical and Mechanical Properties of Red Mud Materials and Stockpile Stability after Dilatation. *Adv. Mater. Sci. Eng.* **2018**, *2018*, 8784232. [CrossRef]
6. Patel, S.; Pal, B.K. Current Status of an Industrial Waste: Red Mud an Overview. *IJLTEMAS* **2015**, *7*, 1–16.
7. Wang, L.; Sun, N.; Tang, H.; Sun, W. A review on comprehensive utilization of red mud and prospect analysis. *Minerals* **2019**, *9*, 362. [CrossRef]
8. Boudreault, R.; Fournier, J.; Primeau, D.; Labrecque-Gilbertm, M.M. Processes for Treating Red Mud. US20150275330A1, 31 January 2017.
9. Xue, S.; Wu, Y.; Li, Y.; Kong, X.; Zhu, F.; William, H.; Li, X.; Ye, Y. Industrial wastes applications for alkalinity regulation in bauxite residue: A comprehensive review. *J. Cent. South Univ.* **2019**, *26*, 268–288. [CrossRef]
10. Arroyo, F.; Luna-Galiano, Y.; Leiva, C.; Vilches, L.F.; Fernandez-Pereira, C. Environmental risks and mechanical evaluation of recycling red mud in bricks. *Environ. Res.* **2020**, *186*, 109537. [CrossRef]
11. Khairul, M.A.; Zanganeh, J.; Moghtaderi, B. The composition, recycling and utilisation of Bayer red mud. *Resour. Conserv. Recyc.* **2019**, *141*, 483–498. [CrossRef]
12. Garg, A.; Yadav, H. Study of red mud as an alternative building material for interlocking block manufacturing in construction industry. *Int. J. Mater. Sci. Eng.* **2015**, *3*, 295–300.
13. Paramguru, R.K.; Rath, P.C.; Misra, V.N. Trends in red mud utilization: A review. *Miner. Process. Extr. Metall. Rev.* **2004**, *26*, 1–29. [CrossRef]
14. Wang, S.; Ang, H.M.; Tade, M.O. Novel applications of red mud as coagulant, adsorbent and catalyst for environmentally benign processes. *Chemosphere* **2008**, *72*, 1621–1635. [CrossRef]
15. Liu, X.; Han, Y.; He, F.; Gao, P.; Yuan, S. Characteristic, hazard and iron recovery technology of red mud—A critical review. *J. Hazard. Mater.* **2021**, *420*, 126542. [CrossRef]
16. Snars, K.; Gilkes, R.J. Evaluation of bauxite residues (red muds) of different origins for environmental applications. *Appl. Clay Sci.* **2009**, *46*, 13–20. [CrossRef]
17. Klauber, C.; Grafe, M.; Power, G. Bauxite residue issues: II. options for residue utilization. *Hydrometallurgy* **2011**, *108*, 11–32. [CrossRef]
18. Piga, L.; Pochetti, F.; Stoppa, L. Recovering metals from red mud generated during alumina production. *JOM* **1993**, *45*, 54–59. [CrossRef]
19. Rai, S.; Bahadure, S.; Chaddha, M.J.; Agnihotri, A. Disposal practices and utilization of red mud (Bauxite Residue): A review in Indian context and abroad. *J. Sustain. Metall.* **2020**, *6*, 1–8. [CrossRef]
20. Das, B.; Mohanty, K. A review on advances in sustainable energy production through various catalytic processes by using catalysts derived from waste red mud. *Renew. Energy* **2019**, *143*, 1791–1811. [CrossRef]
21. Liu, Y.; Lin, C.; Wu, Y. Characterization of red mud derived from a combined Bayer process and bauxite calcination method. *J. Hazard. Mater.* **2007**, *146*, 255–261. [CrossRef]
22. Mayes, W.M.; Jarvis, A.P.; Burke, I.T.; Walton, M.; Feigl, V.R.; Klebercz, O.; Gruiz, K. Dispersal and attenuation of trace contaminants downstream of the Ajka bauxite residue (red mud) depository failure. *Hung. Environ. Sci. Technol.* **2011**, *45*, 5147–5155. [CrossRef] [PubMed]
23. Wang, W.; Pranolo, Y.; Cheng, C.Y. Recovery of scandium from synthetic red mud leach solutions by solvent extraction with D2EHPA. *Sep. Purif. Technol.* **2013**, *108*, 96–102. [CrossRef]
24. Ruyters, S.; Mertens, J.; Vassilieva, E.; Dehandschutter, B.; Poffijn, A.; Smolders, E. The Red Mud Accident in Ajka (Hungary): Plant Toxicity and Trace Metal Bioavailability in Red Mud Contaminated Soil. *Environ. Sci. Technol.* **2011**, *45*, 1616–1622. [CrossRef]
25. Kumar, S.; Kumar, R.; Bandopadhyay, A. Innovative methodologies for the utilisation of wastes from metallurgical and allied industries. *Resour. Conserv. Recycl.* **2006**, *48*, 301–314. [CrossRef]
26. Liu, Z.; Li, H. Metallurgical process for valuable elements recovery from red mud—A review. *Hydrometallurgy* **2015**, *155*, 29–43. [CrossRef]
27. Pascual, J.; Corpas, F.; López-Beceiro, J.; Benítez-Guerrero, M.; Artiaga, R. Thermal characterization of a Spanish red mud. *J. Therm. Anal. Calorim.* **2009**, *96*, 407–412. [CrossRef]
28. Samal, S.; Ray, A.K.; Bandopadhyay, S.A. Proposal for resources, utilization and processes of red mud in India—A review. *Int. J. Miner. Process.* **2013**, *118*, 43–55. [CrossRef]
29. Evans, K. The History, Challenges, and New Developments in the Management and Use of Bauxite Residue. *J. Sustain. Met.* **2016**, *2*, 316–331. [CrossRef]
30. Bonomi, C.; Giannopoulou, I.; Panias, D. Correlation of Scandium and Titanium during Leaching of Bauxite Residue (Red Mud) by an Imidazolium Ionic Liquid. In *Proceedings of the 2nd Conference on European Rare Earth Resources*; Balomenos, E., Marinos, D., Eds.; Heliotopos Conferences Ltd.: Santorini, Greece, 2017; pp. 182–184.
31. Sutar, H.; Mishra, S.C.; Sahoo, S.K.; Maharana, H. Progress of red mud utilization: An overview. *Am. Chem. Sci. J.* **2014**, *4*, 255–279. [CrossRef]
32. Liu, Y.; Naidu, R. Hidden values in bauxite residue (red mud): Recovery of metals. *Waste Manag.* **2014**, *34*, 2662–2673. [CrossRef] [PubMed]

33. Steel industry—Statistics & Facts. Available online: https://www.statista.com/topics/1149/steel-industry/ (accessed on 30 November 2021).
34. Kumar, R.; Srivastava, J.P.; Premchand. Utilization of iron values of red mud for metallurgical applications. In *Environmental and Waste Management*; Bandopadhyay, A., Goswami, N.G., Jamshedpur, P.R.R., Eds.; National Metallurgical laboratories: Jamshedpur, India, 1998; pp. 108–119.
35. Sparks, B.D.; Sirianni, A.F. Beneficiation of a phosphoriferous iron ore by agglomeration methods. *Int. J. Miner. Process.* **1974**, *1*, 231–241. [CrossRef]
36. Bao, Q.; Guo, L.; Guo, Z. A novel direct reduction-flash smelting separation process of treating high phosphorous iron ore fines. *Powder Metall.* **2021**, *377*, 149–162. [CrossRef]
37. Jorgen, D.J. U.S. Geological Survey, Mineral Commodities Summary. Available online: https://minerals.usgs.gov/minerals/pubs/commodity/ironore/mcs-2010-feore.pdf (accessed on 28 November 2020).
38. Muwanguzi, A.J.B.; Karasev, A.V.; Byaruhanga, J.K.; Jönsson, P.G. Characterization of Chemical Composition and Microstructure of Natural Iron Ore from Muko Deposits. *Int. Sch. Res. Notices* **2012**, *2012*, 174803. [CrossRef]
39. Agrawal, A.; Sahu, K.K.; Pandey, B.D. Solid waste management in non-ferrous industries in India. *Resour. Conserv. Recycl.* **2004**, *42*, 99–102. [CrossRef]
40. Abhilash; Sinha, S.; Sinha, M.K.; Pandey, B.D. Extraction of lanthanum and cerium from Indian red mud. *Int. J. Miner. Process.* **2014**, *127*, 70–73. [CrossRef]
41. Mukiza, E.; Zhang, L.; Liu, X.; Zhang, N. Utilization of red mud in road base and subgrade materials: A review. *Resour. Conserv. Recycl.* **2019**, *141*, 187–199. [CrossRef]
42. Pontikes, Y.; Angelopoulos, G.N. Bauxite residue in cement and cementitious applications: Current status and a possible way forward. *Resour. Conserv. Recycl.* **2013**, *73*, 53–63. [CrossRef]
43. Feng, X.-p.; Liu, X.-m.; Sun, H.-h.; Bai, X.; Niu, X.-l. Study on the High Use Ratio of Red Mud in Cementitious Material. *Multipurpose Util. Miner. Resour.* **2007**, *4*, 35–37.
44. Liu, R.-X.; Poon, C.-S. Utilization of red mud derived from bauxite in self-compacting concrete. *J. Clean. Prod.* **2016**, *112*, 384–391. [CrossRef]
45. Nikbin, I.M.; Aliaghazadeh, M.; Sh, C.; Fathollahpour, A. Environmental impacts and mechanical properties of lightweight concrete containing bauxite residue (red mud). *J. Clean. Prod.* **2018**, *172*, 2683–2694. [CrossRef]
46. Kacker, K.P.; Chandra, D. Production of building bricks from red mud rejects of aluminium plants. *Ind. Ceram.* **1977**, *19*, 712–723.
47. Prasad, P.M.; Kachhawha, J.S.; Gupta, R.C.; Mankhand, T.R.; Sharma, J.M. Processing and Applications of Red Muds. In *Key Engineering Materials*; Trans Tech Publications Ltd.: Bäch, Switzerland, 1985; Volume 8, pp. 31–52.
48. Dodoo-Arhin, D.; Konadu, D.S.; Annan, E.; Buabeng, F.P.; Yaya, A.; Agyei-Tuffour, B. Fabrication and Characterisation of Ghanaian Bauxite Red Mud-Clay Composite Bricks for Construction Applications. *Am. J. Mater. Sci.* **2013**, *3*, 110–119.
49. Alam, S.; Das, S.K.; Rao, B.H. Characterization of coarse fraction of red mud as a civil engineering construction material. *J. Clean. Prod.* **2017**, *168*, 679–691. [CrossRef]
50. Wang, L.; Iris, K.M.; Tsang, D.C.W.; Li, S.; Li, J.; Poon, C.S.; Wang, Y.S.; Dai, J.G. Transforming wood waste into water-resistant magnesia-phosphate cement particleboard modified by alumina and red mud. *J. Clean. Prod.* **2017**, *168*, 452–462. [CrossRef]
51. Tang, W.C.; Wang, Z.; Donne, S.W.; Forghani, M.; Liu, Y. Influence of red mud on mechanical and durability performance of self-compacting concrete. *J. Hazard Mater.* **2019**, *379*, 120802. [CrossRef] [PubMed]
52. Kim, Y.; Kim, M.; Sohn, J.; Park, H. Applicability of gold tailings, waste lime stone, red mud, and ferronickel slag for producing glass fibers. *J. Clean. Prod.* **2018**, *203*, 957–965. [CrossRef]
53. Krivenko, P.; Kovalchuk, O.; Pasko, A.; Croymans, T.; Hult, M.; Lutter, G.; Vandevenne, N.; Schreurs, S.; Schroeyers, W. Development of alkali activated cements and concrete mixture design with high volumes of red mud. *Construct. Build. Mater.* **2017**, *151*, 819–826. [CrossRef]
54. Lemougna, P.N.; Wang, K.T.; Tang, Q.; Cui, X.M. Synthesis and characterization of low temperature (<800 °C) ceramics from red mud geopolymer precursor. *Construct. Build. Mater.* **2017**, *131*, 564–573. [CrossRef]
55. Yang, Z.; Mocadlo, R.; Zhao, M.; Sisson, R.D., Jr.; Tao, M.; Liang, J. Preparation of a geopolymer from red mud slurry and class F fly ash and its behavior at elevated temperatures. *Constr. Build. Mat.* **2019**, *221*, 308. [CrossRef]
56. Vigneshwaran, S.; Uthayakumar, M.; Arumugaprabu, V. Potential use of industrial waste-red mud in developing hybrid composites: A waste management approach. *J. Clean. Prod.* **2020**, *276*, 124278. [CrossRef]
57. Amritphale, S.S.; Anshul, A.; Chandra, N.; Ramakrishnan, N. A novel process for making radiopaque materials using bauxite-red mud. *J. Eur. Ceram. Soc.* **2007**, *27*, 1945–1951. [CrossRef]
58. Singh, B.; Gupta, M. Surface treatment of red mud and its influence on the properties of particulate-filled polyester composite. *Bull. Mater. Sci.* **1995**, *18*, 603–621. [CrossRef]
59. Hertel, T.; Cardenia, C.; Balomenos, E.; Panias, D.; Pontikes, Y. Microwave Treatment of Bauxite Residue for the Production of Inorganic Polymers. In Proceedings of the 2nd International Bauxite Residue Valorisation and Best Practices Conference, Athens, Greece, 7–10 May 2018.
60. Qin, S.; Wu, B. Effect of self-glazing on reducing the radioactivity levels of red mud based ceramic materials. *J. Hazard. Mater.* **2011**, *198*, 269–274. [CrossRef] [PubMed]

61. Yang, J.; Zhang, D.; Hou, J.; He, B.; Xiao, B. Preparation of glass-ceramics from red mud in the aluminium industries. *Ceram. Int.* **2008**, *34*, 125–130. [CrossRef]
62. Dry, C.; Meier, J.; Bukowski, J. Sintered coal ash/flux materials for building materials. *Mater. Struct.* **2004**, *37*, 114–121. [CrossRef]
63. Promentilla, M.A.B.; Thang, N.H.; Kien, P.T.; Hinode, H.; Bacani, F.T.; Gallardo, S.M. Optimizing Ternary-blended Geopolymers with Multi-response Surface Analysis. *Waste Biomass Valor.* **2016**, *7*, 929–939. [CrossRef]
64. Hertel, T.; Pontikes, Y. Geopolymers, inorganic polymers, alkali-activated materials and hybrid binders from bauxite residue (red mud)—Putting things in perspective. *J. Clean. Prod.* **2020**, *258*, 120610. [CrossRef]
65. Li, G.H.; Liu, M.X.; Rao, M.J.; Jiang, T.; Zhuang, J.Q.; Zhang, Y.B. Stepwise extraction of valuable components from red mud based on reductive roasting with sodium salts. *J. Hazard. Mater.* **2014**, *280*, 774–780. [CrossRef]
66. Yang, J.K.; Chen, F.; Xiao, B. Engineering application of basic level materials of red mud high level pavement. *China Munic. Eng.* **2006**, *5*, 7–9. [CrossRef]
67. Sushil, S.; Batra, V.S. Catalytic applications of red mud, an aluminium industry waste: A review. *Appl. Catal. B Environ.* **2008**, *81*, 64–77. [CrossRef]
68. Ordonez, S. Comments on catalytic applications of red mud, an aluminium industry waste: A review. *Appl. Catal. B Environ.* **2008**, *84*, 732–733. [CrossRef]
69. Bhatnagar, A.; Vilar, V.J.P.; Botelho, C.M.S.; Boaventura, R.A.R. A review of the use of red mud as adsorbent for the removal of toxic pollutants from water and wastewater. *Environ. Technol.* **2011**, *32*, 231–249. [CrossRef] [PubMed]
70. Ciccu, R.; Ghiani, M.; Serci, A.; Fadda, S.; Peretti, R.; Zucca, A. Heavy metal immobilization in the mining-contaminated soils using various industrial wastes. *Miner. Eng.* **2003**, *16*, 187–192. [CrossRef]
71. Chen, Y.; Li, J.-Q.; Huang, F.; Zhou, J.; Zhou, D.-F.; Liu, W. The performance research on absorbing SO_2 waste gas with Bayer red mud. *J. Guizhou Teachers Univ. Nat. Sci.* **2007**, *36*, 30–32.
72. Brunori, C.; Cremisini, C.; Massanisso, P.; Pinto, V.; Torricelli, L. Reuse of a treated red mud bauxite waste: Studies on environmental compatibility. *J. Hazard. Mater.* **2005**, *117*, 55–63. [CrossRef] [PubMed]
73. Zhang, J.Z.; Yao, Z.Y.; Wang, K.; Wang, F.; Jiang, H.G.; Liang, M.; Wei, J.C.; Airey, G. Sustainable utilization of bauxite residue (Red Mud) as a road material in pavements: A critical review. *Constr. Build Mater.* **2021**, *270*, 121419. [CrossRef]
74. Novais, R.M.; Carvalheiras, J.; Seabra, M.P.; Pullar, R.C.; Labrincha, J.A. Innovative application for bauxite residue: Red mud-based inorganic polymer spheres as pH regulators. *J. Hazard. Mater.* **2018**, *358*, 69–81. [CrossRef] [PubMed]
75. Ujaczki, E.; Feigl, V.; Moln ́ar, M.; Cusack, P.; Curtin, T.; Courtney, R.; O'Donoghue, L.; Davris, P.; Hugi, C.; Evangelou, M.; et al. Re-using bauxite residues: Benefits beyond (critical raw) material recovery. *J. Chem. Technol. Biotechnol.* **2018**, *93*, 2498–2510. [CrossRef]
76. Grafe, M.; Power, G.; Klauber, C. Bauxite residue issues: III. Alkalinity and associated chemistry. *Hydrometallurgy* **2011**, *108*, 60–79. [CrossRef]
77. Narayanan, R.; Kazantzis, P.; Nikolaos, K.; Emmert, M.H. Selective process steps for the recovery of scandium from Jamaican bauxite residue (red mud). *ACS Sustain. Chem. Eng.* **2018**, *6*, 1478–1488. [CrossRef]
78. Guo, D.; Yang, G.; Chen, J.; Fan, Y.; Zhan, H.; Liang, H.; Li, B.; Man, L.; Zhang, J.; Rong, J.; et al. Method for Comprehensively Recovering Scandium and Titanium by Leaching Red Mud with Titanium White Waste Acid. CN103131854B, 28 August 2014.
79. Ochsenkühn-Petropulu, M.; Lyberopulu, T.; Parissakis, G. Direct determination of landthanides, yttrium and scandium in bauxites and red mud from alumina production. *Anal. Chim. Acta* **1994**, *296*, 305–313. [CrossRef]
80. Agrawal, S.; Dhawan, N. Evaluation of red mud as a polymetallic source—A review. *Miner. Eng.* **2021**, *171*, 107084.
81. Vachon, P.; Tyagi, R.D.; Auclair, J.C.; Wilkinson, K.J. Chemical and biological leaching of aluminium from red mud. *Environ. Sci. Technol.* **1994**, *28*, 26–30. [CrossRef]
82. Bruckard, W.J.; Calle, C.M.; Davidson, R.H.; Glenn, A.M.; Jahanshahi, S.; Somerville, M.A.; Sparrow, G.J.; Zhang, L. Smelting of bauxite residue to form a soluble sodium aluminium silicate phase to recover alumina and soda. Miner. Process. *Extract. Metall.* **2010**, *119*, 18–26.
83. Ablamoff, B.; Qian-de, C. *Physical and Chemical Principles of Comprehensive Treatment of Aluminium-containing Raw Materials by Basic Process*; Central South University of Technology Press: Changsha, China, 1988; pp. 178–182.
84. Kannan, P.; Banat, F.; Hasan, S.W.; Haija, M.A. Neutralization of Bayer bauxite residue (red mud) by various brines: A review of chemistry and engineering processes. *Hydrometallurgy* **2021**, *206*, 105758. [CrossRef]
85. Alp, A.; Selim Goral, M. The effects of the additives, calcination and leach conditions for alumina production from red mud. *Scand. J. Metall.* **2003**, *32*, 301–305. [CrossRef]
86. Meher, S.N.; Rout, A.K.; Padhi, B.K. Recovery of Al and Na values from red mud by BaO-Na_2CO_3 sinter process. *Eur. J. Chem.* **2011**, *8*, 1387–1393.
87. Agatzini-Leonardou, S.; Oustadakis, P.; Tsakiridis, P.E.; Markopoulos, C. Titanium leaching from red mud by diluted sulfuric acid at atmospheric pressure. *J. Hazard. Mater.* **2008**, *157*, 579–586. [CrossRef]
88. Erçag, E.; Apak, R. Furnace smelting and extractive metallurgy of red mud: Recovery of TiO2, Al2O3 and pig iron. *J. Chem. Technol. Biotechnol.* **1997**, *70*, 241–246. [CrossRef]
89. Huang, Y.; Chai, W.; Han, G.; Wang, W.; Yang, S.; Liu, J. A perspective of stepwise utilisation of Bayer red mud: Step two—extracting and recovering Ti from Ti enriched tailing with acid leaching and precipitate flotation. *J. Hazard. Mater.* **2016**, *307*, 318–327. [CrossRef]

90. Kasliwal, P.; Sai, P.S.T. Enrichment of titanium dioxide in red mud: A kinetic study. *Hydrometallurgy* **1999**, *53*, 73–87. [CrossRef]
91. Deep, A.; Malik, P.; Gupta, B. Extraction and separation of Ti(IV) using thiophosphinic acids and its recovery from ilmenite and red mud. *Sep. Sci. Technol.* **2001**, *36*, 671–685. [CrossRef]
92. Smirnov, D.I.; Molchanova, T.V. The investigation of sulphuric acid sorption recovery of scandium and uranium from the red mud of alumina production. *Hydrometallurgy* **1997**, *45*, 249–259. [CrossRef]
93. Davris, P.; Balomenos, E.; Panias, D.; Paspaliaris, I. Selective leaching of rare earth elements from bauxite residue (red mud), using a functionalized hydrophobic ionic liquid. *Hydrometallurgy* **2016**, *164*, 125–135. [CrossRef]
94. Lymperopoulou, T.; Georgiou, P.; Tsakanika, L.A.; Hatzilyberis, K.; Ochsenkuehn- Petropoulou, M. Optimizing conditions for scandium extraction from bauxite residue using taguchi methodology. *Minerals* **2019**, *9*, 236. [CrossRef]
95. Zhu, X.; Li, W.; Xing, B.; Zhang, Y. Extraction of scandium from red mud by acid leaching with CaF2 and solvent extraction with P507. *J. Rare Earths.* **2019**, *38*, 1003–1008. [CrossRef]
96. Alkan, G.; Yagmurlu, B.; Cakmakoglu, S.; Hertel, T.; Kaya, S.; Gronen, L.; Stopic, S.; Friedrich, B. Novel Approach for Enhanced Scandium and Titanium Leaching Efficiency from Bauxite Residue with Suppressed Silica Gel Formation. *Sci. Rep.* **2018**, *8*, 5676. [CrossRef] [PubMed]
97. Gu, H.; Li, W.; Li, Z.; Guo, T.; Wen, H.; Wang, N. Leaching Behaviour of Lithium from Bauxite Residue Using Acetic Acid. *Min. Metall. Explor.* **2020**, *37*, 443–451.
98. Panda, S.; Costa, R.B.; Shah, S.S.; Mishra, S.; Bevilaqua, D.; Ata Akcil, A. Biotechnological trends and market impact on the recovery of rare earth elements from bauxite residue (red mud)—A review. *Resour. Conserv. Recycl.* **2021**, *171*, 105645. [CrossRef]
99. Ochsenkuhn-Petropulu, M.; Lyberopulu, T.; Ochsenkuhn, K.M.; Parissakis, G. Recovery of lanthanides and yttrium from red mud by selective leaching. *Anal. Chim. Acta* **1996**, *319*, 249–254. [CrossRef]
100. Ujaczki, É.; Zimmermann, Y.S.; Gasser, C.A.; Molnár, M.; Feigl, V.; Lenz, M. Red mud as secondary source for critical raw materials–extraction study. *J. Chem. Technol. Biotechnol.* **2017**, *92*, 2835–2844. [CrossRef]
101. Shoppert, A.; Loginova, I.; Napol'skikh, J.; Kyrchikov, A.; Chaikin, L.; Rogozhnikov, D.; Valeev, D. Selective Scandium (Sc) Extraction from Bauxite Residue (RedMud Obtained by Alkali Fusion-Leaching Method. *Materials* **2022**, *15*, 433. [CrossRef]
102. Agrawal, S.; Dhawan, N. Investigation of carbothermic microwave reduction followed by acid leaching for recovery of iron and aluminium values from Indian red mud. *Miner. Eng.* **2020**, *159*, 106653. [CrossRef]
103. Chiara, B.; Chiara, C.; Tam, P.; WAI, T.; Dimitrios, P. Review of Technologies in the Recovery of Iron, Aluminium, Titanium and Rare Earth Elements from Bauxite Residue (Red Mud). In Proceedings of the 3rd International Symposium on Enhanced Landfill Mining, Lisbon, Portugal, 8–10 February 2016; pp. 259–276.
104. Stickney, W.A.; Butler, M.O.; Mauser, J.E.; Fursman, O.C. *Utilization of Red Mud Residues from Alumina Production*; US Department of Interior, Bureau of Mines: Washington, DC, USA, 1970.
105. Li, Y.; Wang, J.; Wang, X.; Wang, B.; Luan, Z. Feasibility study of iron mineral separation from red mud by high gradient superconducting magnetic separation. *Phys. C Supercond.* **2011**, *471*, 91–96. [CrossRef]
106. Rai, S.; Nimje, M.T.; Chaddha, M.J.; Modak, S.; Rao, K.R.; Agnihotri, A. Recovery of iron from bauxite residue using advanced separation techniques. *Miner. Eng.* **2019**, *2134*, 222–231. [CrossRef]
107. Borra, C.R.; Blanpain, B.; Pontikes, Y.; Binnemans, K.; Van Gerven, T. Smelting of bauxite residue (red mud) in view of iron and selective rare earths recovery. *J. Sustain. Metall.* **2016**, *22*, 28–37. [CrossRef]
108. Agrawal, S.; Dhawan, N. Microwave carbothermic reduction of low-grade iron ore. Metall. *Mater. Trans. B* **2020**, *51*, 1576–1586. [CrossRef]
109. Shuai, Y.; Xiao, L.; Peng, G.; Yuexin, H. A semi-industrial experiment of suspension magnetization roasting technology for separation of iron minerals from red mud. *J. Hazard. Mater.* **2020**, *394*, 122579.
110. Cardenia, C.; Balomenos, E.; Panias, D. Iron Recovery from Bauxite Residue through Reductive Roastingand Wet Magnetic Separation. *J. Sustain. Metall.* **2019**, *5*, 9–19. [CrossRef]
111. Wanchao, L.; Jiakuan, Y.; Xiao, B. Application of Bayer red mud for iron recovery and building material production from alumosilicate residues. *J. Hazard. Mater.* **2009**, *161*, 474–478.
112. Fanghai, L.; Xiangdong, S.; Fang, H.; Jiawei, W.; Haifeng, W. Co-Treatment of spent pot-lining and red mud for carbon reutilization and recovery of iron, aluminium and sodium by reductive roasting process. *Metall. Mater. Trans. B* **2020**, *51*, 1564–1575.
113. Agrawal, S.; Rayapudi, V.; Dhawan, N. Microwave reduction of red mud for recovery of iron values. *J. Sustain. Metall.* **2018**, *4*, 427–436. [CrossRef]
114. Agrawal, S.; Rayapudi, V.; Dhawan, N. Comparison of microwave and conventional carbothermal reduction of red mud for recovery of iron values. *Miner. Eng.* **2019**, *132*, 202–210. [CrossRef]
115. Javad, S.; Barani, K. Microwave Heating Applications in Mineral Processing. In *The Development and Application of Microwave Heating*; IntechOpen: Rijeka, Croatia, 2012; pp. 79–104.
116. Raspopov, N.A.; Korneev, V.P.; Averin, V.V.; Lainer, Y.A.; Zinoveev, D.V.; Dyubanov, V.G. Reduction of iron oxides during the pyrometallurgical processing of red mud. *Russ. Metall. Met.* **2013**, *1*, 33–37. [CrossRef]
117. Jayasankar, K.; Ray, P.K.; Chaubey, A.K.; Padhi, A.; Satapathy, B.K.; Mukherjee, P.S. Production of pig iron from red mud waste fines using thermal plasma technology. *Int. J. Miner. Metall. Mater.* **2012**, *19*, 679–684. [CrossRef]
118. Zhu, D.Q.; Chun, T.J.; Pan, J.; He, Z. Recovery of iron from high-iron red mud by reduction roasting with adding sodium salt. *Iron Steel Res. Int.* **2012**, *19*, 1–5. [CrossRef]

119. Valeev, D.; Zinoveev, D.; Kondratiev, A.; Lubyanoi, D.; Pankratov, D. Reductive smelting of neutralized red mud for iron recovery and produced pig iron for heat resistant castings. *Metals* **2020**, *10*, 32. [CrossRef]
120. Kaußen, F.; Friedrich, B. Reductive smelting of red mud for iron recovery. *Chem. Ing. Tech.* **2015**, *87*, 1535–1542. [CrossRef]
121. Kongkarat, S.; Khanna, R.; Koshy, O.; O'Kane, P.; Sahajwalla, V. Use of waste Bakelite as a raw material resource for recarburization in steelmaking processes. *Steel Res. Int.* **2011**, *82*, 1228–1239. [CrossRef]
122. Shuai, L.; Zeshuang, K.; Wanchao, L.; Yicheng, L.; Hongshan, Y. Reduction behaviour and direct reduction kinetics of red mud-biomass composite pellets. *J. Sustain. Metall.* **2021**, *7*, 126–135.
123. Rahman, M.; Khanna, R.; Sahajwalla, V.; O'Kane, P. The influence of ash impurities on interfacial reactions between carbonaceous materials and EAF slag at 1550 °C. *ISIJ* **2009**, *49*, 329–336. [CrossRef]
124. Mombelli, D.; Barella, S.; Gruttadauria, A.; Mapelli, C. Iron Recovery from Bauxite Tailings Red Mud by Thermal Reduction with Blast Furnace Sludge. *Appl. Sci.* **2019**, *9*, 4902. [CrossRef]
125. Balomnenos, E.; Kastritis, D.; Panias, D.; Paspaliaris, I.; Boufounos, D. The Enexal Bauxite Residue Treatment Process: Industrial Scale Pilot Plant Results. In *Light Met*; Springer: Cham, Switzerland, 2014; pp. 141–147.
126. Jin, J.; Liu, X.; Yuan, S.; Gao, P.; Li, Y.; Zhang, H.; Meng, X. Innovative utilization of red mud through co-roasting with coal gangue for separation of iron and aluminium minerals. *J. Ind. Eng. Chem.* **2021**, *98*, 298–307. [CrossRef]
127. Cao, Y.; Sun, Y.; Gao, P.; Han, Y.; Li, Y. Mechanism for suspension magnetization roasting of iron ore using straw-type biomass reductant. *Int. J. Min. Sci. Technol.* **2021**, *31*, 1075–1081. [CrossRef]
128. Gostu, S.; Mishra, B.; Martins, G.P. Low Temperature Reduction of Hematite in Red Mud to Magnetite. In *Light Met*; Springer: Cham, Switzerland, 2017; pp. 67–73.
129. Liu, X.; Gao, P.; Yuan, S.; Yang, L.; Han, Y. Clean utilization of high-iron red mud by suspension magnetization roasting. *Miner. Eng.* **2020**, *157*, 106553. [CrossRef]
130. Yu, Z.; Shi, Z.; Chen, Y.; Niu, Y.; Wang, Y.; Wan, P. Red-mud treatment using oxalic acid by UV irradiation assistance. *Trans. Nonferrous Metal. Soc.* **2012**, *222*, 456–460. [CrossRef]
131. Debadatta, D.; Pramanik, K. A study on chemical leaching of iron from red mud using sulphuric acid. *Res. J. Chem. Environ.* **2013**, *17*, 50–56.
132. Pepper, R.A.; Couperthwaite, S.J.; Millar, G.J. Comprehensive examination of acid leaching behaviour of mineral phases from red mud: Recovery of Fe, Al, Ti, and Si. *Miner. Eng.* **2016**, *99*, 8–18. [CrossRef]
133. Yang, Y.; Xuewen, W.; Mingyu, W.; Huaguang, W.; Pengfei, X. Iron recovery from the leached solution of red mud through the application of oxalic acid. *Int. J. Miner. Process.* **2016**, *157*, 145–151. [CrossRef]
134. Eisele, T.C.; Gabby, K.L. Review of reductive leaching of iron by anaerobic bacteria. *Miner. Process. Extractive Metall. Rev.* **2014**, *35*, 75–105. [CrossRef]
135. Laguna, C.; Gonzalez, F.; Garcia-Balboa, C.; Ballester, A.; Blazquez, M.L.; Munoz, J.A. Bio-reduction of iron compounds as a possible clean environmental alternative for metal recovery. *Miner. Eng.* **2011**, *24*, 10–18. [CrossRef]
136. Papassiopi, N.; Vaxevanidou, K.; Paspaliaris, I. Effectiveness of iron reducing bacteria for the removal of iron from bauxite ores. *Miner. Eng.* **2010**, *23*, 25–31. [CrossRef]
137. Vakilchap, F.; Mousavi, S.M.; Shojaosadati, S.A. Role of Aspergillus niger in recovery enhancement of valuable metals from produced red mud in Bayer process. *Bioresour. Technol.* **2016**, *218*, 991–998. [CrossRef]
138. Qu, Y.; Lian, B. Bioleaching of rare earth and radioactive elements from red mud using Penicillium tricolor RM-10. *Bioresour. Technol.* **2013**, *136*, 16–23. [CrossRef] [PubMed]
139. Qu, Y.; Li, H.; Tian, W.; Wang, X.; Wang, X.; Jia, X.; Shi, B.; Song, G.; Tang, Y. Leaching of valuable metals from red mud via batch and continuous processes by using fungi. *Miner. Eng.* **2015**, *81*, 1–4. [CrossRef]
140. Pasechnik, L.A.; Skachkov, V.M.; Chufarov, A.Y.; Suntsov, A.Y.; Yatsenko, S.P. High purity scandium extraction from red mud by novel simple technology. *Hydrometallurgy* **2021**, *202*, 105597. [CrossRef]
141. Khanna, R.; Ellamparuthy, G.; Cayumil, R.; Mishra, S.K.; Mukherjee, P.S. Concentration of Rare Earth Elements during High Temperature Pyrolysis of Waste Printed Circuit Boards. *Waste Manag.* **2018**, *78*, 602–610. [CrossRef]
142. Li, S.; Pan, J.; Zhu, D.; Guo, Z.; Shi, Y.; Dong, T.; Lu, S.; Tian, H. A new route for separation and recovery of Fe, Al and Ti from red mud. *Resour. Conserv. Recycl.* **2021**, *168*, 105314. [CrossRef]
143. Chun, T.; Li, D.; Di, Z.; Long, H.; Tang, L.; Li, F.; Li, Y. Recovery of iron from red mud by high temperature reduction of carbon bearing briquettes. *J. South. Afr. Inst. Min. Metall.* **2017**, *117*, 361–364. [CrossRef]
144. Zhang, Y.; Gao, Q.; Zhao, J.; Li, M.; Qi, Y. Semi-smelting reduction and magnetic separation for the recovery of iron and alumina slag from iron rich bauxite. *Minerals* **2019**, *9*, 223. [CrossRef]
145. Guo, T.; Gao, J.; Xu, H.; Zhao, K.; Shi, X. Nuggets production by direct reduction of high iron red mud. *J. Iron Steel Res. Int.* **2013**, *20*, 24–27. [CrossRef]
146. Samouhos, M.; Taxiarchou, M.; Pilatos, G.; Tsakiridis, P.E.; Devlin, E.; Pissa, M. Controlled reduction of red mud by H_2 followed by magnetic separation. *Miner. Eng.* **2017**, *105*, 36–43. [CrossRef]
147. Cardenia, C.; Balomenos, E.; Wai, P.; Tam, Y.; Panias, D. A Combined Soda Sintering and Microwave Reductive Roasting Process of Bauxite Residue for Iron Recovery. *Minerals* **2021**, *11*, 222. [CrossRef]

148. Mombelli, D.; Mapelli, C.; Barella, S.; Gruttadauria, A.; Ragona, M.; Pisu, M.; Viola, A. Characterization of cast iron and slag produced by red muds reduction via Arc Transferred Plasma (ATP) reactor under different smelting conditions. *J. Environ. Chem. Eng.* **2020**, *8*, 104293. [CrossRef]
149. Maihatchi, A.A.; Pons, M.N.; Ricoux, Q.; Goettmann, F.; Lapicque, F. Production of electrolytic iron from red mud in alkaline media. *J. Environ. Manage.* **2020**, *266*, 110547. [CrossRef] [PubMed]
150. Grudinsky, P.; Zinoveev, D.; Pankratov, D.; Semenov, A.; Panova, M.; Kondratiev, A.; Zakunov, A.; Dyubanov, V.; Petelin, A. Influence of Sodium Sulfate Addition on Iron Grain Growth during Carbothermic Roasting of Red Mud Samples with Different Basicity. *Metals* **2020**, *10*, 1571. [CrossRef]
151. Cheng, C.Y.; Misra, V.N.; Clough, J.; Mun, R. Dephosphorization of Western Australian iron ore by hydrometallurgical process. *Miner. Eng.* **1999**, *12*, 1083–1092. [CrossRef]
152. Srivastava, M.P.; Pan, S.K.; Prasad, N.; Mishra, B.K. Characterization and processing of iron ore fines of Kiruburu deposit of India. *Int. J. Miner. Process.* **2001**, *61*, 93–107. [CrossRef]
153. Guider, J.W. Iron ore beneficiation—key to modern steelmaking. *Min. Eng.* **1981**, *33*, 410–413.
154. Dobbins, M.S.; Burnet, G. Production of an iron ore concentrate from the iron-rich fraction of power plant fly ash. *Resour. Conserv. Recycl.* **1982**, *9*, 231–242. [CrossRef]
155. Worldsteel Association. Available online: https://www.worldsteel.org/media-centre/press-releases/2021/Global-crude-steel-output-decreases-by-0.9--in-2020.html (accessed on 2 December 2021).
156. Zhang, Z.L.; Li, Q.; Zou, Z.S. Reduction properties of high alumina iron ore cold bonded pellet with CO–H_2 mixtures. *Ironmak. Steelmak.* **2014**, *41*, 561–567. [CrossRef]
157. Rath, S.; Sahoo, H.; Dhawan, N.; Rao, D.S.; Das, B.; Mishra, B.K. Optimal Recovery of Iron Values from a Low Grade Iron Ore using Reduction Roasting and Magnetic Separation. *Sep. Sci. Technol.* **2014**, *49*, 1927–1936. [CrossRef]
158. Dhunna, R.; Khanna, R.; Mansuri, I.; Sahajwalla, V. Recycling Waste Bakelite as an Alternative Carbon Resource for Ironmaking Applications. *ISIJ* **2014**, *54*, 613–619. [CrossRef]
159. Kongkarat, S.; Khanna, R.; Koshy, P.; O'kane, P.; Sahajwalla, V. Recycling waste polymers in EAF steelmaking: Influence of polymer composition on carbon/slag interactions. *ISIJ* **2012**, *52*, 385–393. [CrossRef]

MDPI
St. Alban-Anlage 66
4052 Basel
Switzerland
Tel. +41 61 683 77 34
Fax +41 61 302 89 18
www.mdpi.com

Sustainability Editorial Office
E-mail: sustainability@mdpi.com
www.mdpi.com/journal/sustainability

www.ingramcontent.com/pod-product-compliance
Lightning Source LLC
Chambersburg PA
CBHW041955260926
51850CB00019B/2893